Sustain -ability

What to Seek Before Oil Runs Out

Gregory Gebhart

Page 1

Sustain-ability: What to Seek Before Oil Runs Out
Copyright © 2011 Gregory Gebhart
All Rights Reserved

ISBN-13: 978-1461105749
ISB-10: 1461105749

Dedication

This book is dedicated to those who have had the biggest impact on me: my wife, Fran, my parents, Eve and Howard, my brothers, Brian and Eric, and my parents-in-law, Doris and Gary. It is also dedicated to my teachers and professors and to my co-workers and supervisors. This book is also dedicated to those who have had the biggest impact on Sustain-ability: His Royal Highness, the Prince of Wales, former Vice-President Al Gore, and Virgin CEO Sir Richard Branson. Finally it is dedicated to the one true living God.

Introduction

With ever-increasing oil and gasoline prices a person has to be concerned when geologists estimate that oil will be completely gone from our planet in about 50 years. With the increasing civil unrest in the Middle East and the increasing demand for oil from China and India, the upward petroleum price pressures will continue. Because of all of these concerns about the world Petroleum Economy, America and the world need to seek the goal of sustainability. Alternative, renewable, sustainable fuel sources need to be scaled-up immediately. Bio-fuels are the only sustainable transportation fuel source that can be scaled-up right now to replace petroleum and reduce Green House Gas emissions. Bio-fuels are the immediate solution to our economy's need for sustainability, to our need for a Green economy. Sustain-ability is the goal of Earth Day. This book's theme is sustainability.

Chapter 1

Defining Sustain-ability

The United States of America and other countries around the world would benefit from having *SUSTAIN-ABILITY* as the prime goal of their cultures and economies. That is, their main goal should be **SUSTAINING** their citizens, having **SUSTAINABLE** institutions and systems, and having energy sources, farming, and manufacturing with **SUSTAINED YIELDS.** They should provide their citizens with **SUSTENANCE** and **SUSTENATION.** They should always seek SUNSTAIN-ABILITY.

Meriam-Webster's Collegiate Dictionary's 10th edition defines **SUSTAIN** as "to give support or relief to," such as by providing America's adult citizens with adequately paying jobs. It also defines **SUSTAIN** as "to supply with sustenance: NOURISH," such as feeding, clothing, and housing America's citizens through those jobs. Meriam-Webster defines **SUSTAIN** as " to keep UP, PROLONG" and "to support the weight of, to Prop," such as developing vast quantities of renewable energy sources, doing extensive recycling of goods, and promoting highly energy efficient buildings and manufactured products for America's citizens. **SUSTAIN** also means "to buoy up <sustained by hope>," to bear up under," "SUFFER,

UNDERGO, <sustained heavy losses>," and "to support as true, legal or just" such as when entering into a war and seeing that war to its logical conclusion. And *SUSTAIN* also means "to allow or admit as valid <the court sustained the motion>" and "to support by adequate proof: CONFIRM <the testimony sustains our contention>" such as just, logical, and fair judicial decisions would greatly benefit the world's citizens.

Meriam-Webster defines *SUSTAINABLE* as an adjective meaning "of, relating to, or being a method of harvesting or using a resource so that the resource is not depleted or permanently damaged <sustainable techniques> <sustainable agriculture>" such as renewable energy sources and high-energy efficient products and services. The citizens of America and the other countries around the globe would greatly benefit from those countries becoming *SUSTAINABLE*: " of or relating to a lifestyle involving sustainable methods" <sustainable society>."

By developing manufacturing, agriculture, logging, and fishing with *SUSTAINED YIELDS* the citizens of the world would greatly benefit. *SUSTAINABLE YIELD* means "production of a biological resource (as timber or fish) under management procedures which insure replacement of the part harvested by re-growth or reproduction before another harvest occurs."

Providing America's citizens with fine education would give them *SUSTENANCE. SUSTENANCE* is defined as "a means of support, maintenance, or subsistence: LIVING," "FOOD, PROVISIONS, *also* NURISHMENT," "the act of act of sustaining: the state of being sustained," " a supplying or being supplied with the necessaries of life," and "something that gives support, endurance, or strength."

In short people would greatly benefit from their societies providing them with **SUSTENTATION** -a noun meaning "the act of sustaining, the state of being sustained as (a) MAINTENANCE, UPKEEP, (b) PRESERVATION, CONSERVATION, (c) maintenance of life, growth or morale," "something that sustains, SUPPORT."

Everywhere you look in the United States of America and around the World you can see examples of the benefits of societies that sustain themselves and the harms to societies that do not sustain themselves. Whether you look at countries' populations, economies, peace and war, environments, health care systems, farming, manufacturing, education systems, government systems (the laws, their implementation of them, and their courts), or crime rates you see the advantages of countries that take a sustained approach to these phenomena and systems. And you see the disadvantages of non-sustaining societies.

Non-sustaining societies generate and create displaced populations instead of stable populations, unemployment and business failures instead of employment and profits, wars instead of peace; environmental disasters instead of ecologically safe practices, and disease and high health-care costs instead of health and affordable, universal healthcare. Non-sustainable societies are characterized by famines instead of adequate food supplies, shoddy, over-priced, and planned obsolesce products instead of high-quality, reasonably priced, long-lasting products. They have high public school dropout rates and scientifically illiterate students instead of well-educated students, corrupt and incompetent governmental systems instead of honest and competent governments that all of their citizens can be proud of, and high crime rates instead of low-crime through neighborhood-oriented policing.

There is such beauty, simplicity, and elegance of sustainable societal systems: in their economies, in peaceful coexistence, in pollution controls, in health care systems and costs, in farming and adequate food supplies, in high-quality manufactured products, labor and immigration management, in education and literacy, and in government and national morale. Whichever word one uses: ***SUSTAIN, SUSTAINABLE, SUSTAINED YIED, SUSTENANCE, SUSTENATION***, countries throughout the world should be thinking and planning and using the concepts associated with each of these words to inform and direct themselves. In doing so, their citizens will experience the greatest benefits in the greatest numbers.

It is good to point out that as our economies achieve Sustain-ability that they will put behind them the boom-and-bust cycles of the past. No longer will economies fear the inflation of financial bubbles and their inevitable busting. A scale for measuring a business's sustainability would be created by multiplying the percentage of its profits from sustainable activities by its total profits. A scale for measuring a country's sustainability may be created by taking the sum of all of its individual businesses' sustainability scales and dividing them by their total profits.

The glue that holds together **SUSTAINABLE** societies together and which allows them to become **SUSTAINABLE** is ethics and morality. It is no accident that Saxon King Alfred I of the future England began English Jurisprudence in circa 700 AD with a recitation of the 10 Commandments in his legal code, laws that lead to the concept of the King's or Queen's peace.

Chapter 2

2004 Biofuels Proposal

This is a proposal for greatly increasing use of non-fossil fuels by America-for introducing Sustainability to America.

For a while I worked in Houston, Texas at a small refinery that produced LPG, JP-4, Diesel, and Residual Fuel. Part of that job entailed monitoring JP-4 flow into 600,000 barrel tanks at the GATX Tank Farm in the heart of the massive petrochemical complex near the Houston suburb Pasadena. There are a tremendous number of refineries and chemical plants along the Houston Ship Channel that begins as Buffalo Bayou in the center of Houston and runs down to Galveston Bay. Houston is probably the largest nexus of pipelines carrying petroleum, natural gas, petroleum products and chemicals in the United States.

As a teacher of high school physics, I can tell you what any Army, Navy, Marine, or NASA personnel can tell you: using "Newtonian Mechanics": objects sent upward in reaction to the exertion of a force on them (e.g. an artillery charge, a missile booster load, etc.) follow a parabolic path – under the force of gravity – and return to Earth. This principle could be used by enemies of America to fire rockets upward in a parabolic arc landing in the Houston petrochemical complex.

Such an attack would make the 9/11/01 events pale in comparison. The US economy would be smashed. We need to make contingency plans for such a horrible possibility. Petroleum supplies would need to be rerouted around the affected area. Backup fuel generation sources such as wet milling and dry milling corn ethanol production facilities should probably be created for such a contingency (at least 30 million gallons per year so that electricity co-generation could be used to take advantage of the steam used in such facilities.

In the 1973 the US economy was hard hit by loss of much of the oil that it imported through the Arab Oil Embargo. This embargo was imposed as punishment for America's support of our friends, the Israelis, during the Yom Kippur war. This happened during Republican President Gerald Ford's term.

The loss of so much energy from the US economy (at that time energy utilization in dollars made up a bigger part of the Gross Domestic Product then now) sent energy prices skyrocketing. Gas lines formed at filling stations. US Companies passed their higher energy prices through to their customers. This ignited inflation. This inflation was reflected not only in higher commodity costs, but also in higher costs for borrowing money. For example, the interest rates 30 year fixed-rate mortgages rose to almost 18%/year.

There were economists known as "monetarists" who had researched inflation in other countries and they had a solution. Nobel Prize winner and Professor of Economics at the University of Chicago Milton Friedman was their spokesperson. He had researched the German hyperinflation of the 1920s. His recommendation: "hold M1, a key measure of monetary supply, constant." This would halt inflation.

So the FED held money supply measure M1 steady for an extended period of time during President Ronald Reagan's first term by the Board of Governors of the Federal Reserve Banking System in the United States (our central bank, chaired by Paul Volker). The result was inflation dropped dramatically. However, American experienced its worst economic decline since the Great Depression as a direct result.

So if there is another dramatic loss of foreign or domestic energy, not only with the energy itself be far scarcer for transportation, manufacturing, farming, service industry use, and residential use, but prices for the energy as well as other goods and services will skyrocket. Inflation will roar up. The American economy and world economy will falter. There will be massive job losses.

I suggest you do a search of the key words "peak oil" in www.google.com. I especially refer you to the Congressional Testimony of U.S. Congressman Roscoe Bartlett, Ph.D. (Republican) for the 6th District of Maryland. Of course, the Association for the Study of Peak Oil's web site has many individuals such as Texas oilman T. Boone Pickens, university geology and petroleum engineering professors, and energy bankers saying the same unpleasant thing: we need a fall-back energy plan since oil production world wide may have peaked or will shortly max out and domestic natural gas production has already done so.

Combined with the already massive Current Account deficit in the Balance of Trade that will only be worsened by rising oil import prices (China is certainly consuming more and more of this scarce resource), this is a serious problem for business operations to address. The fact that Asian central banks are "financing" both the Trade Deficit and the Fiscal Deficit (federal budget deficit) puts American in an especially awkward financial position. Opening up china to U.S. banking and financial institutions as has just been proposed will help tremendously, but we still need a fall back plan.

On June 11, 2004 Reuters News reported that "the U.S. trade deficit with OPEC had hit a record $5.6 billion with imports from those oil-producing countries at $7.4 billion. The total oil import bill from OPEC and others jumped more than 20 percent in March to a record $10.2 billion....crude oil futures prices [were at] a near-record $40.77 per barrel."

Assuming 12 months per year, that's $120,000,000,000 oil trade deficit per year or 3,000,000,000 barrels/yr oil trade deficit. Assuming 46% of each barrel of imported oil gets refined into gasoline, that's 1,400,000,000 barrels of gasoline/yr from foreign oil or 59,000,000,000 gallons of gasoline/yr.

In 2000 America According to the American Petroleum Institute, in February 2004, the US was importing 12,340,000 barrels/day of oil. Assuming $40/barrel oil and 12,000 barrels of oil imported per day in a 365 day year, that's $175,200,000,000 per year of imported oil or about 25% of the trade deficit or trade gap in 2003 of $489 billion. About 46% of that oil is refined into gasoline or about 2,000,000,000 barrels of gasoline per year in the US is refined from imported oil. Since there are 42 gallons per barrel of gasoline, that means 84,000,000,000 gallons per year of gasoline are refined from foreign oil imports.

Published accounts on the status of the world's oil supply overall hold that it will peak (hit its overall time maximum) shortly or has peaked. Americans are unlikely to drastically change their consumption of oil, oil-refined products (gasoline, diesel, Jet-A, and JP-4, etc.), or oil-derived chemicals and pharmaceuticals. The field of microeconomics gives some clue as to what we can expect when there is decreasing supply of a scarce resource such as oil while demand for that resource is increasing, The price for a scare resource in such a situation invariably increases over a long period of time. As the price for oil trends upward over the long run, Americans will be faced with making personal choices on its use and of products derived from it. There no doubt will be many enterprising individuals and businesses will consider developing and producing alternative energy sources. Assuredly they will also devise ways to use these alternative energy sources in transportation, building climate control, electricity generation, and production of plastics, pharmaceutical, and other chemicals.

It should be pointed out that the case for oil being a "cheap"/"high density energy" source does not seem to consider the fact that 60% of the world's oil is transported by super-tanker. These oil-proponents do not seem to consider that energy necessary to "maintain the speed" (and with it the momentum = Mass * Velocity) of these huge oil tankers over these huge distances. Also most oil in the U.S. and Canada is not "owned" by oil companies (i.e. they do not own the mineral rights, just lease them). A royalty must be paid for each barrel of oil pumped from private U.S. lands, U.S./Canadian government lands (often 1/8th or more). The US Department of Energy has a rough conversion factor of $ to BTUs.

The solution is to do what Brazil (a non-to-small South American country) did during the 1970s while, coincidentally, America's family farmers began to really get hit hard economically: convert over to grain ethanol from gasoline and other petroleum products. Ethanol is the alcohol in beer, wine, and liqueur. By 1987-88, 80% of cars manufactured in Brazil used ethanol for fuel under its Proalcool project. There have been problems with the Proalcool project that mandated the conversion of sugar cane into ethanol.

However, in the article, "Estimating the Net energy Balance of Corn Ethanol," (Agriculture Economic Report No. 721. U.S. Department of Agriculture, Economic Research Service, Office of Energy) by Shapouri and Duffield (Economists with the USDA) and Graboski (Professor of Chemical Engineering at the Colorado School of Mines), "for every BTU (British Thermal Unit) dedicated to producing ethanol (C2H5OH), there is a 24-percent energy gain." The authors also discredited the assumptions of Pimentel's study that concluded that ethanol production for fuel use will result in a net energy loss. The authors go on to say "Producing ethanol from domestic corn stocks achieves a net gain in a more desirable form of energy."

At the web site www.monagbay.com/1013.htm, it is stated "An 85% corn ethanol and 10% unleaded gasoline blend outperforms conventional gasoline and reduces gas emissions by 35-46% while reducing energy use by 50-60%." They authors add "The fuel produces no benzene or sulfur emissions, and very little carbon dioxide and carbon monoxide (when burned as fuel in Brazil's alcohol-powered vehicles [Brazil converted from use of petroleum to national alcohol power in the Proalcool project]; 35% of its emissions are oxygen."

Switching to domestic ethanol as our primary fuel for cars, SUVs, and trucks, would cut way down on pollution from old oil refineries – many of which have been grand-fathered into state pollution laws. This was done time and time again in Texas where my wife and I lived for 23 years.

Shapouri, Duffield, and Graboski also state that "Ethanol production utilizes abundant domestic energy supplies of coal and natural gas to convert corn into a premium liquid fuel that can extend petroleum imports by a factor of 7 to 1." Considering oil is now at about $40/barrel, just think of the positive impact that will have on our country's balance of payments with the other countries in the world that we do business with.

According to the USDA, American farmers harvested corn from about 71,000,000 acres in 2003. Assuming a average yield of 140 bushels of corn per acre as the top nine corn producing states experienced in 2003, that would be 9,940,000,000 bushels of corn. The actual corn production figure for the US in 2003 was 10,000,000,000 bushels. The average conversion ratio of ethanol from corn using wet milling and dry milling is 2.52 gallons of ethanol per bushel of corn. Therefore, the US could produce as much as 25,000,000,000 gallons of ethanol per year if all corn production were diverted to ethanol production.

If the 38,800,000 acres of idle farm land were switched into corn production, this would generate about 5,396,000,000 bushels of corn or 13,490,000,000 additional gallons of ethanol per year.

So we theoretically from corn production alone could generate 38,000,000,000 gallons of ethanol per year or about 64% of a replacement fuel for all foreign oil being imported and converted into gasoline (if one uses the Reuters News figures).

There are several alternative energy sources that may be used in place of oil or oil-derived fuels and chemicals without considerable investment or the development of new technologies for production and use of them (i.e. creating new infrastructures). One of them is ethanol. It may be obtained with significantly greater efficiency from a wide range of natural sources (many of them renewable) than the manner mostly used now to derive ethanol from corn by yeast fermentation (dry- or wet-milling). There exists scientific research and patents for obtaining ethanol without generating the tremendous amounts of carbon dioxide usually associated with traditional dry- or wet-milling processes for obtaining ethanol from corn with fermentation by the yeast Saccharomyces cerevisiae. Of course, this fuel ethanol is the "active ingredient" in all human alcoholic beverages, but before ethanol is used as fuel, it is denatured - made poisonous by adding gasoline to it.

There is research and patents for obtaining ethanol from the corn stover (the corn stalks left behind after the corn is harvested), other grains, grass, paper and wood. And there are many models of cars, trucks, SUVs and minivans that have been produced by the "big three" American auto manufactures with "flexible fuel systems." These "Flexible Fuel Vehicles" (FFV) can accept any mixture of gasoline and ethanol ranging from 90%/10% to 10%/90%. They can effectively burn these mixtures by adjusting their microprocessor-controlled fuel carburetion. Ethanol can also be derived from potatoes and other tubers. Of course grapes, barley, hops, and other traditional alcoholic beverage generation sources may be used to generate ethanol for fuel.

There is also biodiesel that can be derived from soybeans or other legumes. And coal is another viable alternative to oil. Carbon dioxide (CO_2) is a greenhouse gas and is associated with increasing global warming. That is why CO_2 capture and sequestration from electrical power plants have been recommended. But what to do with all of the CO_2? The carbon dioxide generated when coal is burned for heat or to generate electricity can be converted to ethanol by carbon-dioxide-fixing microorganisms (e.g. Acetobacterium woodii). These are some of the short-term alternatives to oil and natural gas. Many U.S. electrical power plants use natural gas a major fuel source. Reliant Energy, the major Houston, TX electric utility, gets about 45% of its electrical power generation capacity from it). Natural gas production in the U.S. has already peaked and is difficult and dangerous to transport by sea in its liquefied state.

While using ethanol, biodiesel, and coal will take the bite out of rising oil prices short term, the use of other energy sources may be expanded: light-water nuclear reactor-produced electricity, solar-generated electricity and heating, wind-generated electricity, hydrodynamic-generated electricity, geothermal-generated electricity and heating, and hydrogen-based fuel.

There are several ways in which to improve upon the current ethanol production techniques from corn and which would allow the use of other plant/tree sources. One is the ZeaChem Process. It uses a two-step fermentation process to produce ethanol from the carbohydrate portions of the plant material. The ZeaChem Process also makes use of the cellulose/hemicellose parts of plant materials to produce ethanol. The ZeaChem Process also produces high-energy-value, single-protein (which may be used as feed for livestock without worries about prions) from the protein portion of the plant materials. No carbon dioxide is produced in the ZeaChem Process. Their patented process envisions converting the acetate salt/acetic acid produced in the second step of the fermentation of the carbohydrate part of the plant material to ethanol using a hyrdrogenation process. This hydrogenation process involves "stripping" hydrogen from methane obtained from either natural gas or from waste reclamation.

There are technologies for converting the carbon dioxide produced by the current ethanol production plants in the United States without reconstructing them to use the ZeaChem Process. There are also technologies for another way to convert the acetate/acetic acid to ethanol than by hydrolysis of it with hydrogen "stripped" from methane by steam. There are ways to even increase the efficiency of existing U.S. ethanol plants by just integrating the fermentation of the corn stover with the fermentation of corn.

The ZiaChem process may be altered by inserting a direct photochemical conversion of the alpha-hydroxyl carboxylic acid, Lactic Acid, produced by the first step of its fermentation process of starches by lactic acid bacteria, to ethanol using semiconductor catalysts. There are certainly a lot of doped and undoped semiconductors that might by adapted to conversion of Lactate/Lactic Acid to Ethanol.

Also, there are microbial processes available to "fix" the carbon dioxide byproduct from conventional commercial yeast fermentation of corn to produce ethanol thereby greatly increasing the efficiency of commercial ethanol production.

There is considerable fallow crop land and considerable grassland/pastures in American farms that may be adapted to provide grasses (cellulose) and grain inputs for commercial ethanol production or legume (soybean) production for biodiesel production commercially.

While the corn ethanol production wouldn't be enough to eliminate all oil imports, it would be enough to eliminate the 2,400,000 barrels per day imports from Saudi Arabia, Iraq and Algeria. (i.e. 876,000,000 barrels per year which is refined into 402,000,000 barrels of gasoline per year or 16,900,000,000 gallons of gasoline per year from those countries.

Converting to domestic ethanol production from petroleum importing and fractional distillation will save our valuable domestic oil reserves for use in being refined into chemical feed stocks. Also farm subsidies to bolster our farmers – family and corporate – in the face of foreign competition could also be greatly reduced as farmland – where practicable – is converted to production of ethanol "agricultural feed stocks." Sound crop rotation policies would have to be employed. And finally, such a change in our policies would directly address as US Senator Pete Dominici's concern in light of terrorist attacks on middle east refineries and oil fields: "This country needs an Energy

Policy. We could be brought to our knees without a single shot being fired."

Also, fuel cell technology in hybrid vehicles could reduce emissions from burning alcohol to just water and oxygen. After doing some "worst case scenario" calculations for the costs of generating ethanol from corn, it was found that this ethanol could be produced for roughly $1.60/gallon (at today's natural gas, electricity, gasoline, diesel, etc. costs). It would be producible for a lot less from dry and wet milling ethanol production plants currently in operation if these prices dropped.

The weighted average of corn prices from the top 9 corn producing states in the Agricultural Statistics 2004 Report from the USDA's ERS, Market and Trade Division, were used to establish this rough figure.
Productivity of almost 1for the US for the past several years gives even more credence to Shapouri, Duffield, and Graboski's determination of a Net Energy Value (NEV) for corn ethanol of +1.24. The productivity figure was used to equate market prices for corn/bushel with total cost of producing the corn/bushel.

This research could be used to revisit their findings from July 1995 under President Bill Clinton and apply current data. On cursory examination of USDA statistics, the present NEV for corn ethanol may be much higher than the authors found. For example, they used the 1990-92 9 State average corn yield of 122 bushels/acre. The unweighted average of yields from corn acreage in the top 9 states is now about 142 bushels/acre.

This is important because the authors used their bushels/acre yields/state to covert energy inputs/acre into BTUs (British Thermal Units) per bushel of the corn going to the milling plants. They were converting BTUs/acre to BTUs/bushel of seed, fertilizer (nitrogen, potash, phosphate, and lime), energy (diesel, gasoline, LPG, electricity, and natural gas), custom work, chemicals, custom drying, and input hauling by dividing their BTUs/acre by bushels/acre. Obviously, the larger the number being divided into the BTUs/acre, the smaller the BTUs/bushel. The NEV for corn ethanol could be more than 5% higher because of this alone.

Also, if the proposed switch to corn ethanol for transportation and heating fuel is adopted, natural gas for electricity generation and oil for chemicals would be freed up. Greater domestic abundance of natural gas and oil would lower prices for electricity, plastics and chemicals that are generated from CH4 (methane) in natural gas. This, in turn, would further raise the NEV of corn ethanol, and lower its cost by the time it's produced at a milling plant.

There are two other big advantages to a shift to large-scale domestic corn ethanol production: it would dramatically drop oil prices, which was a key factor leading to the collapse of the Soviet Union in the 1980s, greatly weakening the influence of OPEC and it would greatly reduce our susceptibility to withdraw of foreign investment to support or federal government deficit and trade deficit. But the shift would have to be made gradually to avoid an "inverse demand shock" on the Net Oil Producing Countries such as OPEC, Russia, Venezuela, Mexico - the mirror image of the "supply shocks" we experienced from the 1973 Arab Oil Embargo.

Another big advantage to a shift to corn ethanol production is that corn, like all plants, consume CO_2 (carbon dioxide – you know, global warming's best friend) and make O_2 (oxygen – man's best friend) from it.

Many scientists are calling for more research on low-cost, cellulosic ethanol production. United States Patent 4,326,032, by Leslie H. Grove of St. Paul, MN should be considered as a cost effective way to do both enzymatic milling and fermentation of corn and corn stover (the stalks left in the field after harvesting the corn). Other cellulosic materials such as switch grass, paper, and wood would be easily adapted to this methodology.

There are ways to further improve the Net Energy Value of biofuels. Grove grew a synergistic mixture of two *Clostridium* (Cl.) bacteria, *Cl. Cellobioparum* and *Cl. Acetobutylicum* (65% to 35%) in anaerobic conditions (90% Nitrogen and 10% Carbon Dioxide) on a growth medium of calf liver and cellulosic material. These two bacteria species grew readily under these conditions.

Then she introduced corncobs and corn stalks (ground up to go through 100 mesh screens) into a fermentation tank (no longer trying to maintain anaerobic conditions) with the growth medium (4$-6% of fermentation tank volume) and heated it to 100 degrees. The resulting fermentation had a yield of roughly 80%. It contained 5% acetone, 11% ethanol, 14% n-butyl alcohol, 41% n-propyl alcohol, 8% n-amyl alcohol and only 1% acetic acid (the remainder (20%) was water).

The heating value of Grove's fuel mix (Example II in her Patent) is 13,412 British Thermal Units (BTU) per pound or about 88,000 BTU per gallon (US) and an octane rating of 114. Ethanol's heating value is 75,700 BTU per gallon (US). Gasoline's low heating value is 115,000 BTU per gallon (US).

Another methodology to examine might be Hess et al's United States Patents 3,212,932 and 3,212,933. They use a two-step, high pressure and temperature (non-biological) process that produces xylose (5 carbon sugar that yeast won't ferment) in the first step and glucose (a six carbon sugar that yeast will ferment) in the second step from lignocellulosic materials. These substrates would be ideal to replace the enzymatic milling of carbohydrates and simultaneous saccharification steps before the first fermentation step of the Zeachem process (US Patent 6,509,180). The Zeachem patent uses *lactobacillus casei* or other monofermentive bacterium that can ferment both sugars. United States Patent 4,515,667 by Rosemary Batroszek-Loza could be used next to photo chemically decarboxylate the lactate produced by the first fermentation step of the Zeachem process. She used a semiconductor titanium oxide as the catalysts to form alcohols like ethanol.

United States Patent 4,461,648 by Patrick Foody allows for increasing the accessibility of cellulose in lignocellulosic materials to chemical and biochemical reagents. Foody and Hess's techniques could be adjusted to replace acid hydrolysis or enzymatic milling of corn used in existing ethanol refineries before fermentation. Also the present ethanol refinery art's production of carbon dioxide

(CO2) from yeast fermentation could be run through an *acetobacterium woodii* medium which would convert most of the CO2 to acetate. Hydrogen (H2) producing bacterium (work well on animal and human waste substrates) could produce the H2 needed to convert the acetate to ethanol (by adding 2 hydrogen atoms per acetate molecule).

Chapter 3
2008
Corn Ethanol & Rain Forest Proposal

This design is for introducing sustainability to America and to the world. It is for removing one billion metric tons of carbon dioxide (CO_2) from the atmosphere in ten years time from today has two fully compatible and highly related components. My design involves the replacement of one billion metric tons of the major transportation petroleum distillate, petrol (gasoline), with a combination of food crop ethanol (from corn or sugar cane grown on active U.S. and Brazilian cropland) and cellulosic ethanol (from switchgrass grown on marginal U.S. cropland) over a ten-year period. The design is economically feasible: in 2006 there were over 5 billion U.S. gallons of corn ethanol produced profitably in the United States and over 5 billion U.S. gallons of sugar cane produced profitably in Brazil. About the same amount of cellulosic ethanol may be annually produced from switchgrass grown on the 19 million acres of marginal U.S. cropland. There have been doubts raised about the economic viability of such ethanol production by Cornell University Professor David Pimentel, however his mathematical model will be shown to be meaningless. There have been doubts raised about the environmental efficacy of food crop ethanol, however these concerns will be shown to not be meaningful. Most importantly it will be shown that using corn and sugar cane ethanol gallon for gallon to replace petrol will result in a 50% reduction in carbon dioxide emissions and replacing petrol with switchgrass ethanol gallon for gallon will result in a 94% reduction in such emissions. There have been ethical questions raised about food crop ethanol: taking food from starving people in third world countries. These ethical concerns will be shown to be misplaced and unfounded fears. Finally, it will also be shown that as the world demand for oil and petroleum distillates increases and as we near "peak oil" (the maximum in worldwide oil production), that having renewable energy sources like food crop ethanol and cellulosic ethanol will become crucial.

Explaining how using biofuels such as food crop ethanol or cellulosic ethanol as fuels will remove carbon dioxide from the atmosphere when they replace petrol involves some stoichiometric calculations (mathematical calculations of mass transfers from chemical reactions). Producing biofuels (with maximum yields) and then burning them (with maximum efficiency) will result in a net reduction in atmospheric carbon dioxide (CO_2), not through removal of existing CO_2 from the atmosphere directly, but indirectly. Plants like corn, sugar cane, and switchgrass sequester carbon dioxide as they grow. Some of this sequestered CO_2 is released in the fermentation process and the rest is released when the biofuels are combusted as fuel. Since corn, sugar cane and switchgrass ethanol are all biofuels, when they are burned instead of petrol, there is no overall net release of CO_2 into the atmosphere. But when petrol (gasoline) is burned as fuel, sequestered CO_2 from plants and animals trapped geologically eons ago is released for the first time, resulting in a net CO_2 release into the atmosphere. The mechanism by which biofuels remove CO_2 from the atmosphere is indirect, yet powerful.

To understand how biofuels will help reduce atmospheric CO_2 consider corn, sugar cane or switchgrass ethanol. The starch in the corn, the disaccharides in sugar cane, the cellulose, hemicellulose, and ligand in switchgrass is formed via biosynthesis (i.e. the combination of photosynthesis and the carbon fixation cycle) according to the following chemical reaction equation:

$6CO_2 + 6H_2O$ + Photons of light with energy (hv) →
$C_6H_{12}O_6$ (glucose) + $6O_2$ (oxygen)

The resulting glucose is polymerized into starch in corn, disaccharides in sugar cane, and cellulose, hemicellulose, and ligand in switchgrass. The corn starch and sugar cane disaccharides are broken down in biorefineries to glucose. This glucose is converted to ethanol by yeast fermentation according to the following chemical reaction equation: $C_6H_{12}O_6$ (glucose) $+2NAD^+ +2ADP +2P_I \rightarrow 2C_2H_5OH$ (ethanol) $+ 2CO_2$.

Said another way; one molecule of glucose produces two molecules of ethanol and two molecules of carbon dioxide. Scaling up the equation using moles (6.023×10^{23} molecules/mole), one mole of glucose produces two moles of ethanol and two moles of carbon dioxide. Cellulosic biorefineries break down the celluose, hemicellose and ligand of switchgrass into sugars like glucose in the saccharification process. This first process in cellosic ethanol production is followed by a second step of fermentation that follow the equation: $C_6H_{12}O_6$ (glucose) $+2NAD^+ +2ADP +2P_I \rightarrow 2C_2H_5OH$ (ethanol) $+ 2CO_2$.

Now is the time for stoichiometric calculations. Ethanol has a density of 0.7894 grams/milliliter at $29^\circ C$ and the conversion factor for gallons into milliliters is 3785.412 according to Lange's Handbook of Chemistry. So one gallon of ethanol has a mass of 3785.412 milliliter (ml) x .7894 gram (gm)/ml = 2988 gm at 29°C. According to Lange's the formula weight of ethanol is 46.07 grams/mole. So there are 2988 gm (ethanol)/46.07 gm/mole (ethanol) = 64.85 moles of ethanol in one gallon of ethanol (= 3.9 x 10^{25} molecules of ethanol in one gallon).

In 2006 the U.S. produced 4.9 billion gallons of corn ethanol and Brazil produced slightly less of sugar cane ethanol. Combined the countries are currently making almost 10 billion gallons of ethanol per year. President George Bush in his January 2007 State of the Union address called for America to begin making 15 billion gallons of ethanol by 2012. If this goal is achieved, America will be producing 15 x 10^9 gallons of ethanol x 64.85 moles/gallon of ethanol = 9.7 x 10^{11} moles of ethanol. For every molecule and every mole of ethanol produced by yeast fermentation, there is one molecule and one mole of carbon dioxide produced, respectively. So when the U.S. is making 15 billion gallons of ethanol/year, 9.7 x 10^{11} moles of carbon dioxide will be produced as a byproduct. This is just keeping track of the CO_2.

Since CO_2 has a formula weight of 44.01 gm/mole, then the 9.7 x 10^{11} moles of CO_2 produced per 9.7 x 10^{11} moles of ethanol refined is equal to 9.7 x 10^{11} moles of CO_2 x 44.01 gm/mole = 4.3 x 10^{13} gm of CO_2. And 4.3 x 10^{13} gm of

CO_2 / 1000 gm/kilogram = 4.3 x 10^{10} kg of CO_2 produced as a byproduct. So 4.3 x 10^{10} kg of CO_2 is produced as a byproduct from refining corn into 15 billion gallons of ethanol. At the current production rate between Brazil and the U.S. of 10 billion gallons of ethanol, 2.9 x 10^{10} kg of CO_2 is the byproduct. This is just keeping track of the CO_2.

But for every mole of glucose in the starch in the corn kernels that produces two moles of ethanol and two moles of carbon dioxide, six moles of carbon dioxide are removed from the atmosphere to make that one mole of glucose. So 3 x 4.3 x 10^{10} kg = 1.3 x 10^{11} kg of CO_2 are removed from the atmosphere to make the glucose that is derived from the starch in the refining of corn kernels into 15 billion gallons of ethanol. That is a net removal of 8.7 x 10^{10} kg of CO_2 (or 1 x 10^{-3} metric tons/kg x 8.7 x 10^{10} kg = 8.7 x 10^7 metric tons of of CO_2) per year from the atmosphere by refining 15 billion gallons of ethanol from corn. So when the United States of America meets its goal of producing 15 billion gallons of ethanol per year, it appears that the U.S. will be removing 87 million metric tons of carbon dioxide per year from the atmosphere. At the current production rate between Brazil and the U.S. of 10 billion gallons of ethanol, 5.7 x 10^{10} kg (57 million metric tons) of CO_2 is being removed from Earth's atmosphere.

But we need to consider that the CO_2 that is produced when ethanol is combusted (when burned as fuel) according to the chemical reaction equation of $2C_2H_5OH$ (ethanol) + $7O_2 \rightarrow 4CO_2 + 6H_2O + O_2$.

For every two moles of ethanol combusted as fuel, four moles of carbon dioxide will be produced. Said another way, for every mole of ethanol produced, 2 moles of carbon dioxide will be produced. We already know that when America produces 15 billion gallons of ethanol it will be equivalent to15 x 10^9 gallons of ethanol x 64.85 moles/gallon of ethanol = 9.7 x 10^{11} moles of ethanol. And when the 9.7 x 10^{11} moles of ethanol are combusted as fuel, 2 x 9.7 x 10^{11} moles of carbon dioxide will be produced or 1.9 x 10^{12} moles of CO_2. At the current production between Brazil and the U.S. of 10 billion gallons of ethanol per year, or 1.2 x 10^{12} moles of CO_2 will be produced as this ethanol is burned as fuel.

Since CO_2 has a formula weight of 44.01 gm/mole, then the 1.9 x 10^{12} moles of CO_2 produced by burning 15 billion gallons of ethanol as fuel is equal to 1.9 x 10^{12} moles of CO_2 x 44.01 gm/mole of CO_2 = 8.4 x 10^{13} gm of CO_2. And 8.4 x 10^{13} gm of CO_2 / 1000 gm/kilogram = 8.4 x 10^{10} kg of CO_2 produced as a when 15 billion gallons of ethanol is burned as fuel and 5.6×10^{10} kg of CO_2 if 10 billion gallons of ethanol per year are burned as fuel.

But as mentioned earlier, for every mole of glucose in the starch in the corn kernels that produces two moles of ethanol and two moles of carbon dioxide, six moles of carbon dioxide are removed from the atmosphere to make that one mole of glucose. So 4.3×10^{10} kg of CO_2 is produced as a byproduct from refining corn into 15 billion gallons of ethanol and 8.4×10^{10} kg of CO_2 is produced as a when ethanol is burned as fuel. But 1.3×10^{11} kg of CO_2 are removed from the atmosphere to make the glucose that is derived from the starch in the refining of corn kernels into ethanol. That is a net removal of zero kg of CO_2 (or zero metric tons of CO_2) per year from the atmosphere by refining 15 billion gallons of ethanol from corn. So when the United States of America meets its goal of producing 15 billion gallons of ethanol per year, it will be removing a net amount of zero metric tons of carbon dioxide per year from the atmosphere. This is the law of conservation of mass. The same is true for burning the current production of 10 billion gallons of ethanol between the U.S. and Brazil.

But for every gallon of gasoline that is replaced by a gallon of corn-derived ethanol, there would be a reduction (as opposed to a removal), a displacement, of carbon dioxide released from the burning of the one-gallon of ethanol as fuel in place of the one gallon of gasoline. So by the earlier stoichiometric calculations and chemical reaction formulas, burning corn or sugar cane ethanol in place of petrol will result in an overall 50% decrease in CO_2 from the displaced petrol released into the atmosphere. According to a January 7, 2008 on-line publication in the *Proceedings of the National Academy of Science* (U.S.) by M.R. Schmer, *et. al.*, "Net energy of cellulosic ethanol from switchgrass," replacing cellulosic switchgrass ethanol with petrol will result in an overall 97% decrease in CO_2 from the displaced petrol released into the atmosphere.

Petrol (gasoline) is composed of about 500 hydrocarbons that vary in length from 5 to 12 carbons. About 25% of petrol (gasoline) is composed of straight chain aliphatic hydrocarbons like heptane (C_7H_{16}). Roughly another 25% of petrol (gasoline) is composed of branched aliphatic hydrocarbons like isooctane (C_8H_{18}). And about another 25% is composed of cyclic aliphatic hydrocarbons like cyclopentane (C_5H_{10}). And again another roughly 25% of petrol (gasoline) is aromatic hydrocarbons like ethyl benzene (C_8H_{10}).

Assuming this crude approximation of the composition of petrol, then one gallon of petrol will contain roughly (0.25 x (((7 x 0.684gm/ml)/ 100.21gm/mole)+((8 x 0.692gm/ml)/ 114.23gm/mole)+((5 x 0.746 gm/ml)/ 70.13gm/mole) +((8 x 0.867 gm/ml)/ 106.17 gm/mole)) x 3785 ml/gallon) moles of carbon per gallon of gasoline. That works out to 203 moles of carbon per gallon of petrol. So if the one-gallon of petrol burns with a 100% yield of carbon dioxide, then roughly 203 moles of CO_2 will be produced. Given a formula weight of carbon dioxide of 44.01 gm/mole, then the 203 moles of CO_2 produced per gallon of petrol burned is equal to 8.92 kg of CO_2 generated when gallon of petrol (gasoline) is burned. Other estimates place the mass of CO_2 from burning one gallon of petrol (gasoline) at 12 kg, but I will use the lower figure of 8.92 kg. So when 15×10^9 gallons of petrol (gasoline) is burned, $1.5 \times 8.9 \times 10^{10}$ kg $(=1.3 \times 10^{11}$ kg) of CO_2 is produced.

If 15 billion gallons of gasoline are not burned as fuel, but is replaced by an equal volume of corn ethanol ethanol annually, then optimally this will reduce the amount of carbon dioxide released into the atmosphere by 1.3×10^8 metric tons of CO_2 per year. That is 130 million metric tons of carbon dioxide that would not be produced by the U.S. per year if 15 billion gallons of petrol (gasoline) is replaced annually by the same number of gallons of ethanol in vehicle fuel tanks. Even at the current American and Brazilian combined ethanol production rate of 10 billion gallons of ethanol per year, this is an annual reduction of 87 million metric tons of carbon dioxide. Both countries are increasing production each year; in 2007, close to 8 billion gallons of ethanol is project by the National Corn Growers Association. Over a ten year period of 2007 to 2017, at current American and Brazilian production rates (atmospheric CO_2 release from internal combustion engines in the United States and Brazil will have decreased by 870 million metric tons. At 15 billion gallons of ethanol production for the U.S. and Brazil, the release of atmospheric carbon dioxide will decline by 1.3 billion metric tons.

Recently two research articles were published in the American Association for the Advancement of Science's (I'm a member) February 29, 2008 *Science*, that advance the notion that corn and sugar cane ethanol produce far more carbon dioxide in its cultivation and harvesting than it saves by sequestering atmospheric carbon dioxide: Joseph Fargione *et al*'s research paper, "Land Clearing and the Biofuel Carbon Debt" and Timothy Searchinger *et al*'s research paper, "Use of U.S. Cropland for Biofuels Increases Greenhouse Gases Through Emissions from Land-Use Change." I agree with their results only insofar are their conclusions apply to *future* burning and tilling of unused farmland to prepare it for corn cultivation or *future* deforestation of land to prepare it for sugarcane or soybean cultivation. Both articles fail to state, as pointed out in an Argon National Laboratories new release offered in rebuttal to them, what their baseline or stating points were for their computer models.

Overall US farmland that was cultivated in 2005 and 2006 for all crops declined by tens of millions of acres over what it was as recently as 1977. In fact all of the corn grown in the US for ethanol production in 2005, 2006 and 2007 was on farmland that had been previously cultivated for corn or soybeans or other crops. The corn for ethanol is being planted on land currently used for farming. By both article's calculations, there would be no "carbon debt" for such corn ethanol. They both are concerned about clearing native vegetation on unused farmland by burning it or cutting it down and leaving it to decompose. Neither article discussed the possibility of converting the native vegetation to cellulosic ethanol immediately upon its removal or sequestering it for later ethanol conversion.

Joseph Fargione *et al*'s research paper, "Land Clearing and the Biofuel Carbon Debt," that appeared in the Febraury 29, 2008 *Science* clearly shows zero carbon debt ($MgCO_2$ ha^{-1}) from prairie biomass ethanol produced from marginal cropland. According to the U.S. Department of Agriculture in 2007 there were 19 million acres of marginal cropland in the United States. There are no crops planted on this marginal cropland, but it is land on which crops, albeit with low yields, could be planted. So if the 19 million acres of US marginal cropland were planted with switch grass, there would be zero carbon debt incurred.

BBC NEWS reported on January 8, 2008 that Ken Vogel *et al*, published findings about bioethanol yields from switch grass (*Panicum virgatum L*) and the resultant carbon dioxide reduction recently in the Proceedings of the National Academy of Science (http://news.bbc.co.uk/go/pr/fr/-/1/hi/sci/tech/7175397.stm). The five year study found that one acre (0.4 hectares) of switch grass could, on average produce 320 gallons of bioethanol. The team also calculated that the production and consumption of switch grass-derived ethanol cut CO2 emissions by about 94% when compared with an equivalent volume of petrol (gasoline). "Greenhouse gas (GHG) emissions of ethanol from switch grass, using only the displacement method, showed 88% less GHG emissions than conventional ethanol," the researchers wrote. In order to maximize the carbon reductions and be economical, Vogel said: "A biorefinery will have a feedstock supply radius of about 25 to 50 miles, so the feedstock of any biorefinery needs to be localized."

So by calculation, we find that planting 19 million acres of marginal US cropland with switch grass and harvesting it annually would, on average, produce 1.9×10^7 acres x 3.20×10^2 gallons of bioethanol/acre = 6.08×10^9 gallons of bioethanol or 6.08 billion gallons of bioethanol per year. And that's without incurring any carbon debt. Also, there would be a reduction in CO2 emissions for using this switch grass ethanol of 94% over petrol (gasoline). Given a formula weight of carbon dioxide of 44.01 gm/mole, then the 203 moles of CO_2 produced per gallon of petrol (gasoline) burned is equal to 8.92 kg of CO_2 generated when a gallon of petrol is burned. And when 6.08×10^9 gallons of petrol is burned, then $6.08 \times 8.9 \times 10^9$ kg (=5.4×10^{10} kg) of CO_2 is produced. Recall that Vogel *et* al found a 94% reduction in CO2 emissions by using switch grass ethanol for fuel instead of petrol. So there would be a $0.94 \times 5.4 \times 10^{10}$ kg = 5.1×10^{10} kg reduction in CO2 produced per year (50 million metric tons) by planting the 19 million acres of marginal US cropland with switch grass, harvesting it, converting it to bioethanol in a biorefinery.

In 2005, according to the US Department of Agriculture, there were 16.7 million acres of US cropland that were allowed to lie fallow during the summer. If those lands were sown with switch grass, harvested in the fall, and converted into bioethanol, that would amount to 1.67×10^7 acres x 3.20×10^2 gallons of bioethanol/acre = 5.3×10^9 gallons of bioethanol per year. There would be zero carbon debt according to Fargione *et al*. When an equivalent amount of petrol (5.3 billion gallons of petrol per year) are combusted, $5.3 \times 8.9 \times 10^9$ kg = 4.8×10^{10} kg of CO_2 are released into the atmosphere. If the switch grass reduced CO_2 emissions 94%, this would be a .94 x = 4.8×10^{10} kg = = 4.5×10^{10} kg reduction in atmospheric CO_2 emissions or a reduction of 45 million metric tons of CO_2 being released into the atmosphere per year. Using some of the millions of acres of US abandoned cropland, such as pastureland for production of switch grass ethanol would only incur a 1-year carbon debt/per acre according to Fargione *et al*.

Besides using large amounts of renewable fuels, another part of establishing a sustainable society is massive recycling. This would include using municipal waste (sewerage) and trash as feedstocks for biofuels. According to Timothy Searchinger *et al*'s research paper, "Use of U.S. Cropland for Biofuels Increases Greenhouse Gases Through Emissions from Land-Use Change," in the February 29, 2008 *Science*, "This study highlights the value of biofuels from waste products because they can avoid land-use changes and its emissions." They continue, "To avoid land-use change altogether, biofuels must use carbon that would reenter the atmosphere without doing useful work that needs to be replaced, for example, municipal waste, crop waste, and fall grass harvests from reserve lands." This echoes what I wrote in my First Addendum to my Virgin Earth Challenge Proposal that neither Fargione's nor Searchinger's research papers address the zero carbon debt approach to converting the biomass cleared from non-cropland to biofuels before corn, sugar cane, soybeans and other biofuel sources are planted. Searchinger also mentions algae grown in the desert as another biofuel feedstock with zero emissions from land-use change.

Jeff Johnson's January 1, 2007 article in the American Chemical Society's *Chemical &Engineering News*, "Ethanol, Is It Worth It?:" contradicts statements that have been made that corn ethanol production would starve third world consumers:

Not much of the U.S. corn crop directly feeds humans, however. About half goes to U.S. animal feed; 20% is exported, mostly to feed foreign animals; 20% makes ethanol; and 10% is used for industrial and some food-related products, around half of which is for high-fructose corn sweeteners. Indeed, corn prices are rising due to ethanol demand, the Department of Agriculture says, and cattle, hogs, poultry, meat, milk, and cheese prices will likely follow suit. These prices will have little effect on the developing world…ethanol production from corn also generates "distillers grains," a valuable high-protein by-product used for cattle feed.

Some parties say that ethanol production would drain huge amounts of potable water from underground aquifers. Unfortunately for them, no water is used directly in the enzymatically catalyzed chemical reaction to produce corn-based ethanol. The glucose in starch in the corn is converted to ethanol by yeast fermentation as follows: one molecule of glucose becomes two molecules of ethanol and two molecules of carbon dioxide. The water is used as a "spectator" only either in creating mixtures or as a solvent in solutions. Aside from water that is lost due to evaporation, most of it should come out in the final step of the process: "The water is filtered through the ground back into the aquifer." In the same way oil refineries use large quantities of water for steam generation to facilitate the fractional distillation of oil into petroleum distillates like gasoline, but no water goes directly into the gasoline hydrocarbon molecules.

Cornell University entomology professor David Pimentel has a mathematical model of all inputs used in corn ethanol production and compared to its energy content when combusted as fuel. Prof. Pimentel concludes that 50% more energy goes into producing corn ethanol than is obtained when it is used as fuel. But Michigan State University chemical engineering professor Bruce Dale applied Pimentel's mathematical model to gasoline and concluded that 100% more energy goes into producing gasoline than it produces as fuel and 250% more energy goes into producing coal than is obtained when it is used as fuel. So Pimentel's model is meaningless.

My response to Professor Pimentel's "Corn Can't Save Us" OP-Ed piece that recently appeared in the St. Louis Post-Dispatch is "Oil and Natural Gas Can't Save Us Either." As oil hovers above $100/barrel, having broken all previous record prices for it in the past two weeks, Americans have to ask themselves what is happening. After being sold record numbers of low-gas mileage trucks and SUVs, we are worried about the recent record prices for gasoline fractionally distilled from oil. 95% of the US transportation system depends on petroleum distallates of gasoline and diesel. 50% of our electrical power comes from coal and 40% from natural gas. The US domestic production of both oil and natural gas peaked (hit maximums) in the last century. Since then they have been in decline. We now import more than 60% of the oil we use, often from politically highly unstable areas.Our trade deficit, which we must borrow money to finance, is at record levels because of these imports. Inflation is rising because of the steep increase in the price per barrel of oil. Even food prices are sky-rocketing. Worse still, the US Energy Information Administartion says in roughly 20 years, the world-wide production of oil will peak and go into irreversible decline. Corn ethanol is not the only alternative energy source available to the US. There are coal, nuclear, ocean and river current trubines, solar, wind, geothermal, hydroelectric, fuels obtained from human trash and waste waste, ect. No single alternative energy source will replace oil or natural gas, but time is running out. By the way, Professor Pimentel's mathematical model also arrives at the conclusion that gasoline takes 100% more energy to produce than is obtained when it is used as a fuel compared to ethanol's 46%.

Chapter 4

2008 Proposal for Replacement of Fossil Fuels and Coal with Biofuels

The catastrophic problem of global warming arose largely from a pattern of global industrialization through cheap, non-renewable, non-sustainable energy sources such as petroleum distillates (petrol (gasoline), Jet-A, JP-4, diesel, etc.) and coal, and lack of recycling. One long-term solution to the global warming and its harmful climatic changes would be to displace the petroleum distillates with either biofuels generated from starch saccharification and saccharide fermentation to ethanol or biofuels generated through bacterial conversion of carbon dioxide (CO_2) and methane (CH_4) emissions of coal-fueled electrical generation. Then greatly increased recycling of paper, aluminum, plastic, and glass would be added. Nations that switch to sustainable energy sources in transportation, reduce CO_2 emissions in electricity generation, and which reduce paper, aluminum, plastics and glass production energy uses and CO_2 through recycling would drastically alter the CO_2 emissions causing global warming. Sustainability is the key.

In 2006 the United States of America contributed over 5.89 billion metric tons of the world's CO_2 emissions. Those American emissions come largely through petroleum distillate transportation fuel combustion and coal-fueled electricity generation. Over 1.186 billion metric tons of CO_2 emissions in 2006 came from US gasoline (petrol). Over 1.937 billion metric tons of CO_2 emissions in 2006 came from US electricity generated by coal combustion. These contribute about 50% of the US CO_2 emissions. So this third Virgin Earth Challenge Design of mine will concentrate on displacement of petroleum distillate fuels with biofuels from food crops such as corn and sugar cane (through starch saccharification and saccharide fermenetation by yeast) and with biofuels from ethanol co-generation plants built at coal-fired electricity generation facilities (instead of CO_2 sequestration underground) to remove one billion metric tons of carbon dioxide (CO_2) from the atmosphere in ten years time from today.

My design involves the replacement of about ten billion metric gallons of the major transportation petroleum distillate, petrol (gasoline), with a combination of food crop ethanol (from corn or sugar cane grown on active U.S. and Brazilian cropland) and cellulosic ethanol (from switchgrass grown on marginal U.S. cropland) over a ten-year period.

The design is economically feasible: in 2006 there were over 5 billion U.S. gallons of corn ethanol produced profitably in the United States and over 5 billion U.S. gallons of sugar cane produced profitably in Brazil. About the same amount of cellulosic ethanol may be annually produced from switchgrass grown on the 19 million acres of marginal U.S. cropland. There have been doubts raised about the economic viability of such ethanol production by Cornell University Professor David Pimentel, however his mathematical model will be shown to be meaningless. There have been doubts raised about the environmental efficacy of food crop ethanol, however these concerns will be shown to not be meaningful. Most importantly it will be shown that using corn and sugar cane ethanol gallon for gallon to replace petrol will result in a 50% reduction in carbon dioxide emissions and replacing petrol with switchgrass ethanol gallon for gallon will result in a 94% reduction in such emissions. There have been ethical questions raised about food crop ethanol: taking food from starving people in third world countries. These ethical concerns will be shown to be misplaced and unfounded fears.

Explaining how using biofuels such as food crop ethanol or cellulosic ethanol as fuels will remove carbon dioxide from the atmosphere when they replace petrol involves some stoichiometric calculations (mathematical calculations of mass transfers from chemical reactions). Producing biofuels (with maximum yields) and then burning them (with maximum efficiency) will result in a net reduction in atmospheric carbon dioxide (CO_2), not through removal of existing CO_2 from the atmosphere directly, but indirectly. Plants like corn, sugar cane, and switchgrass sequester carbon dioxide as they grow. Some of this sequestered CO_2 is released in the fermentation process and the rest is released when the biofuels are combusted as fuel. Since corn, sugar cane and switchgrass ethanol are all biofuels, when they are burned instead of petrol, there is no overall net release of CO_2 into the atmosphere. But when petrol (gasoline) is burned as fuel, sequestered CO_2 from plants and animals trapped geologically eons ago is released for the first time, resulting in a net CO_2 release into the atmosphere. The mechanism by which biofuels remove CO_2 from the atmosphere is indirect, yet powerful.

Likewise taking global warming causing coal CO_2 and CH_4 emissions in electricity generation and creating ethanol with them via genetically modified bacteria patented by University of Oklahoma scientists. A company has been set up to take advantage of this process by converting the CO_2 and CH_4 from biomass to ethanol which has been invested in by General Motors, but it can be applied to coal emissions as well. If the coal CO_2 and CH_4 emissions are not released into the atmosphere but are converted into ethanol which gallon per gallon, liter for liter, replaces gasoline (petrol) and with that displacement harmful CO_2 emissions from the gasoline (petrol). This decreases the net CO_2 from coal-fired electricity generation and transportation fuel combustion.

To understand how biofuels from corn and sugar cane will help reduce atmospheric CO_2 consider corn, sugar cane or switchgrass ethanol. The starch in the corn, the disaccharides in sugar cane, the cellulose, hemicellulose, and ligand in switchgrass is formed via biosynthesis (i.e. the combination of photosynthesis and the carbon fixation cycle) according to the following chemical reaction equation:

$6CO_2 + 6H_2O$ + Photons of light with energy (hv) $\rightarrow$
$C_6H_{12}O_6$ (glucose) + $6O_2$ (oxygen)

The resulting glucose is polymerized into starch in corn, disaccharides in sugar cane, and cellulose, hemicellulose, and ligand in switchgrass. The corn starch and sugar cane disaccharides are broken down in biorefineries to glucose. This glucose is converted to ethanol by yeast fermentation according to the following chemical reaction

equation: $C_6H_{12}O_6$ (glucose) $+2NAD^+ +2ADP +2P_I$ →
$2C_2H_5OH$ (ethanol) + $2CO_2$.

Said another way; one molecule of glucose produces two molecules of ethanol and two molecules of carbon dioxide. Scaling up the equation using moles (6.023×10^{23} molecules/mole), one mole of glucose produces two moles of ethanol and two moles of carbon dioxide. Cellulosic biorefineries break down the celluose, hemicellose and ligand of switchgrass into sugars like glucose in the saccharification process. This first process in cellosic ethanol production is followed by a second step of fermentation that follow the equation: $C_6H_{12}O_6$ (glucose) $+2NAD^+ +2ADP +2P_I$ → $2C_2H_5OH$ (ethanol) + $2CO_2$.

Now is the time for stoichiometric calculations. Ethanol has a density of 0.7894 grams/milliliter at 29°C and the conversion factor for gallons into milliliters is 3785.412 according to Lange's Handbook of Chemistry. So one gallon of ethanol has a mass of 3785.412 milliliter (ml) x .7894 gram (gm)/ml = 2988 gm at 29°C. According to Lange's the formula weight of ethanol is 46.07 grams/mole. So there are 2988 gm (ethanol)/46.07 gm/mole (ethanol) = 64.85 moles of ethanol in one gallon of ethanol (= 3.9×10^{25} molecules of ethanol in one gallon).

In 2006 the U.S. produced 4.9 billion gallons of corn ethanol and Brazil produced slightly less of sugar cane ethanol. Combined the countries are currently making almost 10 billion gallons of ethanol per year. President George Bush in his January 2007 State of the Union address called for America to begin making 15 billion gallons of ethanol by 2012. If this goal is achieved, America will be producing 15 x 10^9 gallons of ethanol x 64.85 moles/gallon of ethanol = 9.7 x 10^{11} moles of ethanol. For every molecule and every mole of ethanol produced by yeast fermentation, there is one molecule and one mole of carbon dioxide produced, respectively. So when the U.S. is making 15 billion gallons of ethanol/year, 9.7 x 10^{11} moles of carbon dioxide will be produced as a byproduct. This is just keeping track of the CO_2.

Since CO_2 has a formula weight of 44.01 gm/mole, then the 9.7 x 10^{11} moles of CO_2 produced per 9.7 x 10^{11} moles of ethanol refined is equal to 9.7 x 10^{11} moles of CO_2 x 44.01 gm/mole = 4.3 x 10^{13} gm of CO_2. And 4.3 x 10^{13} gm of CO_2 / 1000 gm/kilogram = 4.3 x 10^{10} kg of CO_2 produced as a byproduct. So 4.3 x 10^{10} kg of CO_2 is produced as a byproduct from refining corn into 15 billion gallons of ethanol. At the current production rate between Brazil and the U.S. of 10 billion gallons of ethanol, 2.9×10^{10} kg of CO_2 is the byproduct. This is just keeping track of the CO_2.

But for every mole of glucose in the starch in the corn kernels that produces two moles of ethanol and two moles of carbon dioxide, six moles of carbon dioxide are removed from the atmosphere to make that one mole of glucose. So 3 $\times 4.3 \times 10^{10}$ kg $= 1.3 \times 10^{11}$ kg of CO_2 are removed from the atmosphere to make the glucose that is derived from the starch in the refining of corn kernels into 15 billion gallons of ethanol. That is a net removal of 8.7×10^{10} kg of CO_2 (or 1×10^{-3} metric tons/kg $\times 8.7 \times 10^{10}$ kg $= 8.7 \times 10^{7}$ metric tons of of CO_2) per year from the atmosphere by refining 15 billion gallons of ethanol from corn. So when the United States of America meets its goal of producing 15 billion gallons of ethanol per year, it appears that the U.S. will be removing 87 million metric tons of carbon dioxide per year from the atmosphere. At the current production rate between Brazil and the U.S. of 10 billion gallons of ethanol, 5.7×10^{10} kg (57 million metric tons) of CO_2 is being removed from Earth's atmosphere.

But we need to consider that the CO_2 that is produced when ethanol is combusted (when burned as fuel) according to the chemical reaction equation of $2C_2H_5OH$ (ethanol) $+ 7O_2 \rightarrow 4CO_2 + 6H_2O + O_2$.

For every two moles of ethanol combusted as fuel, four moles of carbon dioxide will be produced. Said another way, for every mole of ethanol produced, 2 moles of carbon dioxide will be produced. We already know that when America produces 15 billion gallons of ethanol it will be equivalent to 15 x 10^9 gallons of ethanol x 64.85 moles/gallon of ethanol = 9.7 x 10^{11} moles of ethanol. And when the 9.7 x 10^{11} moles of ethanol are combusted as fuel, 2 x 9.7 x 10^{11} moles of carbon dioxide will be produced or 1.9 x 10^{12} moles of CO_2. At the current production between Brazil and the U.S. of 10 billion gallons of ethanol per year, or 1.2 x 10^{12} moles of CO_2 will be produced as this ethanol is burned as fuel.

Since CO_2 has a formula weight of 44.01 gm/mole, then the 1.9 x 10^{12} moles of CO_2 produced by burning 15 billion gallons of ethanol as fuel is equal to 1.9 x 10^{12} moles of CO_2 x 44.01 gm/mole of CO_2 = 8.4 x 10^{13} gm of CO_2. And 8.4 x 10^{13} gm of CO_2 / 1000 gm/kilogram = 8.4 x 10^{10} kg of CO_2 produced as a when 15 billion gallons of ethanol is burned as fuel and 5.6 x 10^{10} kg of CO_2 if 10 billion gallons of ethanol per year are burned as fuel.

But as mentioned earlier, for every mole of glucose in the starch in the corn kernels that produces two moles of ethanol and two moles of carbon dioxide, six moles of carbon dioxide are removed from the atmosphere to make that one mole of glucose. So 4.3 x 10^{10} kg of CO_2 is produced as a byproduct from refining corn into 15 billion gallons of ethanol and 8.4 x 10^{10} kg of CO_2 is produced as a when ethanol is burned as fuel. But 1.3 x 10^{11} kg of CO_2 are removed from the atmosphere to make the glucose that is derived from the starch in the refining of corn kernels into ethanol. That is a net removal of zero kg of CO_2 (or zero metric tons of CO_2) per year from the atmosphere by refining 15 billion gallons of ethanol from corn. So when the United States of America meets its goal of producing 15 billion gallons of ethanol per year, it will be removing a net amount of zero metric tons of carbon dioxide per year from the atmosphere. This is the law of conservation of mass. The same is true for burning the current production of 10 billion gallons of ethanol between the U.S. and Brazil.

But for every gallon of gasoline that is replaced by a gallon of corn-derived ethanol, there would be a reduction (as opposed to a removal), a displacement, of carbon dioxide released from the burning of the one-gallon of ethanol as fuel in place of the one gallon of gasoline. So by the earlier stoichiometric calculations and chemical reaction formulas, burning corn or sugar cane ethanol in place of petrol will result in an overall 50% decrease in CO_2 from the displaced petrol released into the atmosphere. According to a January 7, 2008 on-line publication in the *Proceedings of the National Academy of Science* (U.S.) by M.R. Schmer, *et. al.*, "Net energy of cellulosic ethanol from switchgrass," replacing cellulosic switchgrass ethanol with petrol will result in an overall 97% decrease in CO_2 from the displaced petrol released into the atmosphere.

Petrol (gasoline) is composed of about 500 hydrocarbons that vary in length from 5 to 12 carbons. About 25% of petrol (gasoline) is composed of straight chain aliphatic hydrocarbons like heptane (C_7H_{16}). Roughly another 25% of petrol (gasoline) is composed of branched aliphatic hydrocarbons like isooctane (C_8H_{18}). And about another 25% is composed of cyclic aliphatic hydrocarbons like cyclopentane (C_5H_{10}). And again another roughly 25% of petrol (gasoline) is aromatic hydrocarbons like ethyl benzene (C_8H_{10}).

Assuming this crude approximation of the composition of petrol, then one gallon of petrol will contain roughly (0.25 x (((7 x 0.684gm/ml)/ 100.21gm/mole)+((8 x 0.692gm/ml)/ 114.23gm/mole)+((5 x 0.746 gm/ml)/ 70.13gm/mole) +((8 x 0.867 gm/ml)/ 106.17 gm/mole)) x 3785 ml/gallon) moles of carbon per gallon of gasoline. That works out to 203 moles of carbon per gallon of petrol. So if the one-gallon of petrol burns with a 100% yield of carbon dioxide, then roughly 203 moles of CO_2 will be produced. Given a formula weight of carbon dioxide of 44.01 gm/mole, then the 203 moles of CO_2 produced per gallon of petrol burned is equal to 8.92 kg of CO_2 generated when gallon of petrol (gasoline) is burned. Other estimates place the mass of CO_2 from burning one gallon of petrol (gasoline) at 12 kg, but I will use the lower figure of 8.92 kg. So when 15×10^9 gallons of petrol (gasoline) is burned, $1.5 \times 8.9 \times 10^{10}$ kg $(=1.3 \times 10^{11}$ kg) of CO_2 is produced.

If 15 billion gallons of gasoline are not burned as fuel, but is replaced by an equal volume of corn ethanol annually, then optimally this will reduce the amount of carbon dioxide released into the atmosphere by 1.3×10^8 metric tons of CO_2 per year. That is 130 million metric tons of carbon dioxide that would not be produced by the U.S. per year if 15 billion gallons of petrol (gasoline) is replaced annually by the same number of gallons of ethanol in vehicle fuel tanks. Even at the current American and Brazilian combined ethanol production rate of 10 billion gallons of ethanol per year, this is an annual reduction of 87 million metric tons of carbon dioxide. Both countries are increasing production each year; in 2007, close to 8 billion gallons of ethanol is projected by the National Corn Growers Association. Over a ten year period of 2007 to 2017, at current American and Brazilian production rates, atmospheric CO_2 release from internal combustion engines in the United States and Brazil will have decreased by 870 million metric tons. At 15 billion gallons of ethanol production for the U.S. and Brazil, the release of atmospheric carbon dioxide will decline by 1.3 billion metric tons.

Recently two research articles were published in the American Association for the Advancement of Science's (I'm a member) February 29, 2008 *Science*, that advance the notion that corn and sugar cane ethanol produce far more carbon dioxide in its cultivation and harvesting than it saves by sequestering atmospheric carbon dioxide: Joseph Fargione *et al*'s research paper, "Land Clearing and the Biofuel Carbon Debt" and Timothy Searchinger *et al*'s research paper, "Use of U.S. Cropland for Biofuels Increases Greenhouse Gases Through Emissions from Land-Use Change." I agree with their results only insofar are their conclusions apply to *future* burning and tilling of unused farmland to prepare it for corn cultivation or *future* deforestation of land to prepare it for sugarcane or soybean cultivation. Both articles fail to state, as pointed out in an Argon National Laboratories new release offered in rebuttal to them, what their baseline or stating points were for their computer models.

Overall US farmland that was cultivated in 2005 and 2006 for all crops declined by tens of millions of acres over what it was as recently as 1977. In fact all of the corn grown in the US for ethanol production in 2005, 2006 and 2007 was on farmland that had been previously cultivated for corn or soybeans or other crops. The corn for ethanol is being planted on land currently used for farming. By both article's calculations, there would be no "carbon debt" for such corn ethanol. They both are concerned about clearing native vegetation on unused farmland by burning it or cutting it down and leaving it to decompose. Neither article discussed the possibility of converting the native vegetation to cellulosic ethanol immediately upon its removal or sequestering it for later ethanol conversion.

Joseph Fargione *et al*'s research paper, "Land Clearing and the Biofuel Carbon Debt," that appeared in the Febraury 29, 2008 *Science* clearly shows zero carbon debt ($MgCO_2$ ha^{-1}) from prairie biomass ethanol produced from marginal cropland. According to the U.S. Department of Agriculture in 2007 there were 19 million acres of marginal cropland in the United States. There are no crops planted on this marginal cropland, but it is land on which crops, albeit with low yields, could be planted. So if the 19 million acres of US marginal cropland were planted with switch grass, there would be zero carbon debt incurred.

BBC NEWS reported on January 8, 2008 that Ken Vogel *et al*, published findings about bioethanol yields from switch grass (*Panicum virgatum L*) and the resultant carbon dioxide reduction recently in the Proceedings of the National Academy of Science (http://news.bbc.co.uk/go/pr/fr/-/1/hi/sci/tech/7175397.stm). The five year study found that one acre (0.4 hectares) of switch grass could, on average produce 320 gallons of bioethanol. The team also calculated that the production and consumption of switch grass-derived ethanol cut CO2 emissions by about 94% when compared with an equivalent volume of petrol (gasoline). "Greenhouse gas (GHG) emissions of ethanol from switch grass, using only the displacement method, showed 88% less GHG emissions than conventional ethanol," the researchers wrote. In order to maximize the carbon reductions and be economical, Vogel said: "A biorefinery will have a feedstock supply radius of about 25 to 50 miles, so the feedstock of any biorefinery needs to be localized."

So by calculation, we find that planting 19 million acres of marginal US cropland with switch grass and harvesting it annually would, on average, produce 1.9×10^7

acres x 3.20 x 10^2 gallons of bioethanol/acre = 6.08 x 10^9 gallons of bioethanol or 6.08 billion gallons of bioethanol per year. And that's without incurring any carbon debt. Also, there would be a reduction in CO2 emissions for using this switch grass ethanol of 94% over petrol (gasoline). **Given a formula weight of carbon dioxide of 44.01 gm/mole, then the 203 moles of CO_2 produced per gallon of petrol (gasoline) burned is equal to 8.92 kg of CO_2 generated when a gallon of petrol is burned. And when 6.08 x 10^9 gallons of petrol is burned, then 6.08 x 8.9 x 10^9 kg (=5.4 x 10^{10} kg) of CO_2 is produced. Recall that Vogel *et* al found a 94% reduction in CO2 emissions by using switch grass ethanol for fuel instead of petrol. So there would be a 0.94 x 5.4 x 10^{10} kg = 5.1 x 10^{10} kg reduction in CO2 produced per year (50 million metric tons) by planting the 19 million acres of marginal US cropland with switch grass, harvesting it, converting it to bioethanol in a biorefinery.**

In 2005, according to the US Department of Agriculture, there were 16.7 million acres of US cropland that were allowed to lie fallow during the summer. If those lands were sown with switch grass, harvested in the fall, and converted into bioethanol, that would amount to 1.67 x 10^7 acres x 3.20 x 10^2 gallons of bioethanol/acre = 5.3 x 10^9 gallons of bioethanol per year. There would be zero carbon debt according to Fargione *et al*. When an equivalent amount of petrol (5.3 billion gallons of petrol per year) are combusted, $5.3 \times 8.9 \times 10^9$ kg $= 4.8 \times 10^{10}$ kg of CO_2 are released into the atmosphere. If the switch grass reduced CO_2 emissions 94%, this would be a $.94 \times = 4.8 \times 10^{10}$ kg $= 4.5 \times 10^{10}$ kg reduction in atmospheric CO_2 emissions or a reduction of 45 million metric tons of CO_2 being released into the atmosphere per year. Using some of the millions of acres of US abandoned cropland, such as pastureland for production of switch grass ethanol would only incur a 1-year carbon debt/per acre according to Fargione *et al*.

Besides using large amounts of renewable fuels, another part of establishing a sustainable society is massive recycling. This would include using municipal waste (sewerage) and trash as feedstocks for biofuels. According to Timothy Searchinger *et al*'s research paper, "Use of U.S. Cropland for Biofuels Increases Greenhouse Gases Through Emissions from Land-Use Change," in the February 29, 2008 *Science*, "This study highlights the value of biofuels from waste products because they can avoid land-use changes and its emissions." They continue, "To avoid land-use change altogether, biofuels must use carbon that would reenter the atmosphere without doing useful work that needs to be replaced, for example, municipal waste, crop waste, and fall grass harvests from reserve lands." Neither Fargione's nor Searchinger's research papers address the zero carbon debt approach to converting the biomass cleared from non-cropland to biofuels before corn, sugar cane, soybeans and other biofuel sources are planted. Searchinger also mentions algae grown in the desert as another biofuel feedstock with zero emissions from land-use change.

Jeff Johnson's January 1, 2007 article in the American Chemical Society's *Chemical &Engineering News*, "Ethanol, Is It Worth It?:" contradicts statements that have been made that corn ethanol production would starve third world consumers:

Not much of the U.S. corn crop directly feeds humans, however. About half goes to U.S. animal feed; 20% is exported, mostly to feed foreign animals; 20% makes ethanol; and 10% is used for industrial and some food-related products, around half of which is for high-fructose corn sweeteners. Indeed, corn prices are rising due to ethanol demand, the Department of Agriculture says, and cattle, hogs, poultry, meat, milk, and cheese prices will likely follow suit. These prices will have little effect on the developing world...ethanol production from corn also generates "distillers grains," a valuable high-protein by-product used for cattle feed.

Some parties say that ethanol production would drain huge amounts of potable water from underground aquifers. Unfortunately for them, no water is used directly in the enzymatically catalyzed chemical reaction to produce corn-based ethanol. The glucose in starch in the corn is converted to ethanol by yeast fermentation as follows: one molecule of glucose becomes two molecules of ethanol and two molecules of carbon dioxide. The water is used as a "spectator" only either in creating mixtures or as a solvent in solutions. Aside from water that is lost due to evaporation, most of it should come out in the final step of the process: "The water is filtered through the ground back into the aquifer." In the same way oil refineries use large quantities of water for steam generation to facilitate the fractional distillation of oil into petroleum distillates like gasoline, but no water goes directly into the gasoline hydrocarbon molecules.

Cornell University entomology professor David Pimentel has a mathematical model of all inputs used in corn ethanol production and compared to its energy content when combusted as fuel. Prof. Pimentel concludes that 50% more energy goes into producing corn ethanol than is obtained when it is used as fuel. But Michigan State University chemical engineering professor Bruce Dale applied

Pimentel's mathematical model to gasoline and concluded that 100% more energy goes into producing gasoline than it produces as fuel and 250% more energy goes into producing coal than is obtained when it is used as fuel. So Pimentel's model is meaningless.

When one considers the energy inputs to create either food corn or corn ethanol, one discovers from the Cornell University mathematical model being touted by entomology Professor David Pimentel that more energy goes into producing either than is the energy found in the end results: food corn or corn ethanol. So food corn, for either human or animal consumption, requires huge amounts of energy to grow, harvest, and turn into food. And from a regression analysis of corn energy input prices against corn market prices it is clear that there is a very high correlation between the two. This means that most increases in corn energy input prices are passed directly through to consumers of the food corn.

Food corn and corn ethanol energy inputs are found in fuels, fertilizers, pesticides, farm machinery, and others. Fuels for farming are mostly diesel and gasoline (petrol). Diesel prices and gasoline (petrol) prices have gone up in almost complete synchronization with crude oil prices. According to the U.S. Energy Information Administration, the Freight-On-Board spot price for crude oil from all countries (Petroleum Navigator) went from January 2006 to $54.63/barrel in January 2007 to $92.93/barrel in January 2008.to $116/barrel in May 2008. Interesting the price received by Minnesota farmers for corn in January of 2004 was $2.24/bushel, in January 2005 was $1.94/bushel, in January 2006 was $1.74/bushel, and January 2007 was $2.94. U.S. farmers received $5.01/ bushel in March 2007 and $6.00/bushel in May 2008 for corn.

So the price of crude oil increased by 70.1 % from January of 2006 to January 2007 and the price of corn increased by 69.0 % from January 2006 to January 2007. And the price of crude oil increased by 24.8% from January 2008 to May 2008 while the price of corn increased by 20% in the U.S. between March of 2007 and March of 2008. So when oil prices rise 1%, corn prices rise almost 1% too.

This raises the question are higher corn prices being caused by corn ethanol production, by increased international demand for corn, by energy prices, or by some combination of all three. Since all three have increased from 2007 until 2008, it clearly is a combination of all three. How much is energy involved? Look at the main individual inputs to corn: fuel, fertilizer, pesticides, supplies, farm machinery and labor. According to the US Department of Agriculture, fuel costs for farmers increased 200% from 2002 to 2005. Fertilizer costs increased 61% for farmers over the same period. The others increased only between 10% and 15% over the same period. Fuel had been about 13% of the total operating costs (input costs) for corn farmers in 2006, but this will have increased by 22% of total corn farmer operating costs by 2007 and 31.8% by May of 2008. And energy intensive Ammonium nitrate fertilizer costs have also been skyrocketing.

It is clear: higher energy costs are the main cause of higher corn prices. This input-output model explains why farmers of other crops such as wheat and soybeans have also seen higher operating costs. It explains much of those crops' higher prices. So corn ethanol is not the major contributor to higher food prices, rising energy costs are.

Chapter 5

Biofuels & Hydrokinetic Energy Design

Design Title: "Sustainable Energy Reduces CO2 Emissions"

Design Inventor: Gregory Gebhart

Submission Date: January 4, 2009

Abstract of Design:

This Design substitutes sustainable energy for fossil fuels to reduce carbon dioxide (CO2) in the Earth's atmosphere. The Design does so by eliminating or replacing emissions from combustion of fossil fuels of crude oil and its distillates, natural gas, and coal. The substitution of fresh and seawater for fossil fuels eliminates fossil fuel emissions. Substituting ethanol derived from biomass for fossil fuel distillates substitutes emissions from the former with emissions from the latter. In another embodiment of the design, this substitution may be accomplished with ethanol created from coal combustion emissions.

References (This Design Incorporates the following References):

Footnotes:

[1] Bedard, Roger, et al. North American Ocean Energy Status – March 2007. 2007. Proceedings of the 7th European

Wave and Tidal Energy Conference. 11-13 September 2007. Porto, Portugal. Calculations include 260TWh of wave-generated electricity and 140 TWh from tidal and in-stream electricity. The estimates cited in the Proceedings assume a 15% conversion rate of hydrokinetic energy to mechanical energy, power train efficiencies and conversion availability of 90%. Our calculation assumes electricity use of 6,000 kWh per year for a typical non-electric heating U.S. household.

[2] Dixon, Douglas. EPRI. "The Future of Waterpower: 23,000 MW+ by 2025." June 2007. Environment and Energy Study Institute briefing. Washington, DC. And Personal communication, R. Bedard, EPRI. April 2008. Online at: http://www.hydro.org/hydrofacts/EPRIEESITheFutureof Waterpower060807.pdf

[3] Assumes an average new coal plant generating capacity of 600 MW.

[4] Assumes a heat rate of 8,870 Btu/kWh for a new supercritical pulverized coal plant based on MIT data (Future of Coal, 2007), a carbon content for coal of 220 lbs/million Btu based on EIA data, and tailpipe emissions of 12,100 lbs/year for an average car based on EPA data.

Bibliography:

M.R. Schmer, K.P. Vogel, R.B. Mitchell. And R.K. Perrin, "Net Energy of Cellulosic Ethanol from Switchgrass," *Proceedings of the National Academy of Sciences*, January 15, 2008, Vol. 105, No. 2, pages 494-469. Appearing at

http://www.pnas.org/content/105/2/464.full

http://www.cnbc.com/id/24875687/ - "Clean Coal Technologies"

http://www.xconomy.com/boston/2008/08/21/coskata-refutes-energy-analysts-critique-says-its-on-track-to-make-ethanol-for-under-1-per-gallon/

http://www.sgm.ac.uk/news/consultations/sgmcon054.pdf - "Different Biofuel Syntheses"

Claims of the Design:

In one embodiment of the Design, ethanol production from corn or sugar cane and substitution of this ethanol for gasoline will economically remove over one billion tons of atmospheric CO_2 over the next ten years.

In one embodiment of the Design, ethanol production from switch grass or a mixture of prairie grasses and substitution of this ethanol for gasoline will economically remove over one billion tons of atmospheric CO_2 over the next ten years

One embodiment of the Design, ethanol generation from coal combustion emissions will economically make 100 gallons of ethanol from one ton of coal. If all coal-fired electrical generation facilities in the U.S. are equipped to convert coal emissions to ethanol, and the ethanol substituted for gasoline, this will remove one billion tons of atmospheric CO2 over ten years.

One embodiment of the Design is the generation of electricity from the kinetic energy of water (river and ocean currents or wave motion). **The amount of energy that can be captured from U.S. waves, tides, and river currents is estimated to be enough to power over 67 million homes.** [1] Based on current hydrokinetic proposals, it is predicted that America could be producing 13,000 MW of power from hydrokinetic energy by 2025. [2] **This is equivalent to displacing 22 new dirty coal-fired power plants** [3] - eliminating the annual emission of nearly 86 million metric tons of carbon dioxide, as well as other harmful pollutants like mercury and particulate matter. The eliminated carbon emissions in 2025 would be equivalent to taking 15.6 million cars off the road. [4]

Description of the Design:

One embodiment of the Design is Bio-ethanol production from corn and sugar cane. Plants like corn and sugar cane are grown in the soil using photosynthesis, carbon fixation (CO_2 uptake and conversion into sugars), water, and minerals are the source material for the Biomass used in bio-ethanol production. These plants' high starch parts are ground up and combined with water to make Mash. Microorganisms like yeast and E. coil ferment the Mash to make Bio-ethanol and release carbon dioxide. The plants' high cellulose and hemi-cellulose parts are reduced to complex sugars in a saccharification process. Then those saccharides (sugars) are converted to Bio-ethanol by microorganisms like poly-fermentative bacteria.

Another embodiment of the Design is Bio-ethanol production from switch grass or a mixture of prairie grasses. Using the Coskata process, the switch grass or a mixture of fast-growing grasses is vaporized in gasifiers to Syngas rich in H_2 (hydrogen) and CO_2 (carbon dioxide). The Syngas is passed through pipes lined with membranes coated with bacteria that have been genetically engineered by scientists at the University of Oklahoma to ethanol. Then the ethanol is washed away by water.

Another embodiment of the Design is ethanol production from coal combustion emissions. Using the "clean coal" technology of Integrated Gasification Combined Cycle (IGCC), coal is converted in a gasifier at high temperatures, in the presence of a metal catalyst, to Syngas. Syngas is composed of CO_2, CH_4 (methane) and H_2 (hydrogen). In IGCC electrical generation, this Syngas is combusted to generate electricity in a combustion turbine. But there are excess H_2, CO_2, and CH_4 left over from the IGCC gasifier and combustion turbine. Further electrical generation may be made from these leftovers. But the coal emissions can be converted by genetically engineered bacteria into ethanol. This is a Fischer-Tropsch process developed by Coskata.

Another embodiment of the Design is electrical generation by wave motion, eliminating fossil fuel emissions with energy from the motion of water. Buoy's are set in the ocean or large lakes. These buoys have electrical generation wiring to convert the up and down motion of magnets into current running through coiled copper wiring running around the magnets.

Another embodiment of the Design is electrical generation by ocean or river currents, eliminating fossil fuel emissions with energy from the motion of water. Water turbines are set below the water surface in the ocean and in rivers. The movement of sea or fresh water (ocean and river currents) drives these turbines and electricity is generated.

Summary of the Design:

One of the preferred embodiment of the Design uses M.R. Schmer *et al*'s published findings about Bio-ethanol yields from switch grass (*Panicum virgatum L*) and the resultant carbon dioxide reduction recently in the Proceedings of the National Academy of Science. Schmer's five year study found that one hectare (0.4 hectares) of switch grass could, on average produce an average of 8.4×10^3 kg of switch grass biomass. This would be $0.4 \times 8.4 \times 10^3$ kg = 3.4×10^3 kg per acre of switch grass biomass.

So in this preferred embodiment of the Design, we find that planting 19 million acres of marginal US cropland with switch grass and harvesting it annually would, on average, produce 1.9×10^7 acres x 3.4×10^3 kg per acre of switch grass biomass = 6.5×10^{10} kg of switch grass biomass per year (6.5×10^7 metric tons). And in this preferred embodiment of the Design, the Fischer-Tropsch Caskata process yields 100 gallons of Bio-ethanol per metric ton of switch grass biomass. So 19 million acres of marginal crop land planted with switch grass would yield 6.5×10^7 metric tons of switch grass biomass x 10^2 gallons of ethanol per metric ton of biomass per year or 6.5 billion gallons of ethanol per year. These 6.5 billion gallons of ethanol will replace roughly the same volume of petrol. This would make a significant reduction in CO2 emissions.

Given a formula weight of carbon dioxide of 44.01 gm/mole, then the 203 moles of CO_2 produced per gallon of petrol (gasoline) burned is equal to 8.92 kg of CO_2 generated when a gallon of petrol is burned. And when 6.5×10^9 gallons of petrol is burned, then $6.5 \times 8.9 \times 10^9$ kg (=5.8×10^{10} kg) of CO_2 is produced. In this preferred embodiment of the Design, the Coskata process would be substituted for the saccharification and fermentation of Schmer's Bio-Refinery. This would result in an 84% reduction in CO2 emissions compared to those from gasoline combustion. So there would be a $0.84 \times 5.8 \times 10^{10}$ kg = 4.9×10^{10} kg reduction in CO2 produced per year (49 million metric tons) by planting the 19 million acres of marginal US cropland with switch grass, harvesting it, converting it to Bio-ethanol in the Fischer-Tropsch Coscata process. The embodiment of the Design allows for about 100 gallons of ethanol per one ton of material feedstock and less than one gallon of water per gallon of ethanol produced. Using corn ethanol requires three-to-seven times as much water for every gallon of ethanol produced.

Another preferred embodiment of the Design relies on a 12-year study University of Minnesota, Twin Cities, ecologist David Tillman was able to attain twice as much ethanol from plots of mixtures of prairie grasses (switch grass mixed with 15 native perennial grasses) than from plots with only switch grass. Tillman's Bio-ethanol production with a mixture of prairie grasses was a 238% higher than has been obtained with corn. He did this without irrigated or fertilizing the sites with these mixtures of prairie grasses.

This other preferred embodiment of the design the mixed prairie grass plots produced twice as much ethanol as Schimer's all switch grass plots. So if 19 million acres of marginal crop land planted with the 15 native perennial grasses, this acreage would yield 6.5×10^7 metric tons of switch grass biomass x 2×10^2 gallons of ethanol per metric ton of biomass per year or 13 billion gallons of ethanol per year in the Coskata process. These 13 billion gallons of ethanol will replace roughly the same volume of petrol. This would make a significant reduction in CO2 emissions.

Given a formula weight of carbon dioxide of 44.01 gm/mole, then the 203 moles of CO_2 produced per gallon of petrol (gasoline) burned is equal to 8.92 kg of CO_2 generated when a gallon of petrol is burned. And when 13×10^9 gallons of petrol is burned, then $13 \times 8.9 \times 10^9$ kg (=1.2×10^{11} kg) of

CO_2 is produced. In this other preferred embodiment of the Design, the Coskata process would be substituted for the saccharification and fermentation of Schmer's Bio-Refinery. This would result in an 84% reduction in CO2 emissions compared to those from gasoline combustion. So there would be a $0.84 \times 1.2 \times 10^{11}$ kg $= 1.0 \times 10^{11}$ kg reduction in CO2 produced per year (100 million metric tons) by planting the 19 million acres of marginal US cropland with 15 native perrenial grasses, harvesting them, converting them to Bio-ethanol in the Fischer-Tropsch Coskata process. The embodiment of the Design allows for about 100 gallons of ethanol per one ton of material feedstock and less than one gallon of water per gallon of ethanol produced. Using corn ethanol requires three-to-seven times as much water for every gallon of ethanol produced.

Chapter 6
Switch Grass Biofuels Proposal

Design Title: "Attaining Sustainability by Substituting
Ethanol for Fossil Fuels Reduces CO2 Emissions"
Design Inventor: Gregory Gebhart
Submission Date: January 9, 2009

Abstract of Design:

This Design substitutes sustainable energy for fossil
fuels to reduce carbon dioxide (CO2) in the Earth's
atmosphere. The Design does so by eliminating or
replacing emissions from combustion of fossil fuels of crude
oil and its distillates, natural gas, and coal. Substituting
ethanol derived from biomass for fossil fuel distillates
substitutes emissions from the former with emissions from
the latter. In another embodiment of the design, this
substitution may be accomplished with ethanol created from
coal combustion emissions.

References (This Design Incorporates the following
References):
M.R. Schmer, K.P. Vogel, R.B. Mitchell. And R.K. Perrin,
"Net Energy of Cellulosic Ethanol from Switchgrass,"

Proceedings of the National Academy of Sciences, January 15, 2008, Vol. 105, No. 2, pages 494-469. Appearing at
http://www.pnas.org/content/105/2/464.full
http://www.xconomy.com/boston/2008/08/21/coskata-refutes-energy-analysts-critique-says-its-on-track-to-make-ethanol-for-under-1-per-gallon/
http://www.sgm.ac.uk/news/consultations/sgmcon054.pdf -
"Different Biofuel Syntheses"

Claims of the Design:

In one embodiment of the Design, ethanol production from corn or sugar cane and substitution of this ethanol for gasoline will economically remove over one billion tons of atmospheric CO_2 over the next ten years.

In one embodiment of the Design, ethanol production from switch grass or a mixture of prairie grasses and substitution of this ethanol for gasoline will economically remove over one billion tons of atmospheric CO_2 over the next ten years.

Description of the Design:

One embodiment of the Design is Bio-ethanol production from corn and sugar cane. Plants like corn and sugar cane are grown in the soil using photosynthesis, carbon fixation (CO_2 uptake and conversion into sugars), water, and minerals are the source material for the Biomass used in bio-ethanol production. These plants' high starch parts are ground up and combined with water to make Mash. Microorganisms like yeast and E. coil ferment the Mash to make Bio-ethanol and release carbon dioxide. The plants' high cellulose and hemi-cellulose parts are reduced to complex sugars in a saccharification process. Then those saccharides (sugars) are converted to Bio-ethanol by microorganisms like poly-fermentative bacteria.

Another embodiment of the Design is Bio-ethanol production from switch grass or a mixture of prairie grasses. Using the Coskata process, the switch grass or a mixture of fast-growing grasses is vaporized in gasifiers to Syngas rich in H_2 (hydrogen) and CO_2 (carbon dioxide). The Syngas is passed through pipes lined with membranes coated with bacteria that have been genetically engineered by scientists at the University of Oklahoma to ethanol. Then the ethanol is washed away by water.

Summary of the Design:

One of the preferred embodiment of the Design uses M.R. Schmer *et al*'s published findings about Bio-ethanol yields from switch grass (*Panicum virgatum L*) and the resultant carbon dioxide reduction recently in the Proceedings of the National Academy of Science. Schmer's five year study found that one hectare (0.4 hectares) of switch grass could, on average produce an average of 8.4×10^3 kg of switch grass biomass. This would be $0.4 \times 8.4 \times 10^3$ kg = 3.4×10^3 kg per acre of switch grass biomass.

So in this preferred embodiment of the Design, we find that planting 19 million acres of marginal US cropland with switch grass and harvesting it annually would, on average, produce 1.9×10^7 acres x **3.4 x 10^3 kg per acre of switch grass biomass** = 6.5×10^{10} kg of switch grass biomass per year (6.5×10^7 metric tons). And in this preferred embodiment of the Design, the Fischer-Tropsch Caskata process yields 100 gallons of Bio-ethanol per metric ton of switch grass biomass. So 19 million acres of marginal crop land planted with switch grass would yield 6.5×10^7 metric tons of switch grass biomass x 10^2 gallons of ethanol per metric ton of biomass per year or 6.5 billion gallons of ethanol per year. These 6.5 billion gallons of ethanol will replace roughly the same volume of petrol. This would bake a significant reduction in CO2 emissions.

Given a formula weight of carbon dioxide of 44.01 gm/mole, then the 203 moles of CO_2 produced per gallon of petrol (gasoline) burned is equal to 8.92 kg of CO_2 generated when a gallon of petrol is burned. And when 6.5 x 10^9 gallons of petrol is burned, then 6.5 x 8.9 x 10^9 kg (=5.8 x 10^{10} kg) of CO_2 is produced. In this preferred embodiment of the Design, the Coskata process would be substituted for the saccharification and fermentation of Schmer's Bio-Refinery. This would result in an 84% reduction in CO2 emissions compared to those from gasoline combustion. So there would be a 0.84 x 5.8 x 10^{10} kg = 4.9 x 10^{10} kg reduction in CO2 produced per year (49 million metric tons) by planting the 19 million acres of marginal US cropland with switch grass, harvesting it, converting it to Bio-ethanol in the Fischer-Tropsch Coscata process. The embodiment of the Design allows for about 100 gallons of ethanol per one ton of material feedstock and less than one gallon of water per gallon of ethanol produced. Using corn ethanol requires three-to-seven times as much water for every gallon of ethanol produced.

Another preferred embodiment of the Design relies on a 12-year study University of Minnesota, Twin Cities, ecologist David Tillman was able to attain twice as much ethanol from plots of mixtures of prairie grasses (switch grass mixed with 15 native perennial grasses) than from plots with only switch grass. Tillman's Bio-ethanol production with a mixture of prairie grasses was a 238% higher than has been obtained with corn. He did this without irrigated or fertilizing the sites with these mixtures of prairie grasses.

This other preferred embodiment of the design the mixed prairie grass plots produced twice as much ethanol as Schimer's all switch grass plots. So if 19 million acres of marginal crop land planted with the 15 native perennial grasses, this acreage would yield 6.5×10^7 metric tons of switch grass biomass x 2×10^2 gallons of ethanol per metric ton of biomass per year or 13 billion gallons of ethanol per year in the Coskata process. These 13 billion gallons of ethanol will replace roughly the same volume of petrol. This would bake a significant reduction in CO_2 emissions. Given a formula weight of carbon dioxide of 44.01 gm/mole, then the 203 moles of CO_2 produced per gallon of petrol (gasoline) burned is equal to 8.92 kg of CO_2 generated when a gallon of petrol is burned. And when 13×10^9 gallons of petrol is burned, then $13 \times 8.9 \times 10^9$ kg ($=1.2 \times 10^{11}$ kg) of CO_2 is produced. In this other preferred embodiment of the Design, the Coskata process would be substituted for the saccharification and fermentation of Schmer's Bio-Refinery. This would result in an 84% reduction in CO_2 emissions compared to those from gasoline combustion. So there would be a $0.84 \times 1.2 \times 10^{11}$ kg $= 1.0 \times 10^{11}$ kg reduction in CO_2

produced per year (100 million metric tons) by planting the 19 million acres of marginal US cropland with 15 native perrenial grasses, harvesting them, converting them to Bio-ethanol in the Fischer-Tropsch Coskata process. The embodiment of the Design allows for about 100 gallons of ethanol per one ton of material feedstock and less than one gallon of water per gallon of ethanol produced. Using corn ethanol requires three-to-seven times as much water for every gallon of ethanol produced.

Chapter 7

Recycling Engine Emissions Proposal

Design: "Attaining Sustainability by Recycling Fossil Fuels
to Reduce CO2 Emissions"

Designer: Gregory Gebhart

Date: March 10, 2009

Background of the Design:

The recycling of materials such as paper, plastics, and aluminum has recently grown into a major enterprise around the world. But there is much more recycling that could be done. Recycling of fossil fuels is a major area that needs to be exploited if carbon dioxide emissions are to be reigned in and global warming reversed. Such broad scale recycling will change un-sustainable energy sources into sustainable ones. This extensive recycling will become a key characteristic of a sustainable society.

Specifically, the emissions from the combustion of petrol (gasoline), diesel, Jet-A, and JP-4 fossil fuels may be recycled and made into more petrol, diesel, Jet_A and JP-4. Jet-A is aviation fuel for private propellar and jet aircraft amd JP-4 is military aviation fuel. To recycle these fuels' emissions back into their original fuels, there vehicles and aircraft will have to be hybridized by adding small chemical cells to them to accomplish this recycling.

Summary of the Design:

One embodiment of the Design is the recycling of petrol combustion emissions. The recycling cell for gasoline may be poisoned and deactivated by the presence of sulfur in the petrol and sulfur dioxide in the petrol emissions. In one embodiment of this design, sulfur-binding agents like those made of iron will be added to the petrol so the iron in these additives binds with the sulfur in the petrol. When the petrol is combusted, there will be almost no sufur dioxide created.

Ferox is one such agent. To prevent the iron in these additives from contaminating the recycled petrol, before the petrol is combusted, the petrol will go through a baffle chamber with magnetized fan blades rotating very fast in staggered positions along a straight path. The sulfur bound to the iron will adhere to the blades. When the petrol is cobusted, its emissions will pass through a catlytic converter which will convert hydrocarbons to carbon dioxide and water and nitrous oxide to nitrogen and oxygen.The carbon dioxide and water will pass through a recycling cell that will convert them into petrol and water using the Fischer-Tropsch process. The recycled petrol will go back into the fuel supply for the vehicle.

One embodiment of the Design is the recycling of diesel combustion emissions. The recycling cell for diesel may be poisoned and deactivated by the presence of sulfur in the diesel and sulfur dioxide in the diesel emissions. In one embodiment of this design, sulfur-binding agents like those made of iron will be added to the diesel so the iron in these additives binds with the sulfur in the diesel. When the diesel is combusted, there will be almost no sulfur dioxide created. Ferox is one such agent. To prevent the iron in these additives from contaminating the recycled diesel, before the diesel is combusted, the diesel will go through a baffle chamber with magnetized fan blades rotating very fast in staggered positions along a straight path. The sulfur bound to the iron will adhere to the blades.

When the diesel is combusted, its emissions will pass through a catalytic converter which will convert hydrocarbons to carbon dioxide and water and nitrous oxide to nitrogen and oxygen. The carbon dioxide and water will pass through a recycling cell that will convert them into diesel and water using the Fischer-Tropsch process. The diesel petrol will go back into the fuel supply for the vehicle.

One embodiment of the Design is the recycling of Jet-A combustion emissions. The recycling cell for Jet-A may be poisoned and deactivated by the presence of sulfur in the Jet-A and sulfur dioxide in the Jet-A emissions. In one embodiment of this design, sulfur-binding agents like those made of iron will be added to the Jet-A so the iron in these additives binds with the sulfur in the Jet-A. When the Jet-A is combusted, there will be almost no sulfur dioxide created. Ferox is one such agent. To prevent the iron in these additives from contaminating the recycled Jet-A, before the Jet-A is combusted, the Jet-A will go through a baffle chamber with magnetized fan blades rotating very fast in staggered positions along a straight path. The sulfur bound to the iron will adhere to the blades. When the Jet-A is combusted, its emissions will pass through a catalytic converter which will convert hydrocarbons to carbon dioxide and water and nitrous oxide to nitrogen and oxygen. The carbon dioxide and water will pass through a recycling cell that will convert them into Jet-A and water using the Fischer-Tropsch process. The recycled Jet-A will go back into the fuel supply for the vehicle.

Chapter 8

Converting Atmospheric CO2 to Biofuels Proposal
Design Name: "Attaining Sustainability by Large Scale
Fixation of CO2 From Air and Conversion into Synthetic
Petrol"
Designer: Gregory Gebhart
Design Submission Date; April 16, 2009

Design Summary:

This design economically removes large quantities of
carbon dioxide (CO_2) from the Earth's atmosphere. It is
a four step process. It starts with CO_2 and ends up with
synthetic petrol (gasoline) and diesel fuel. The first step
is to remove large quantities of CO_2 from the air [1].

The second step is to desalinate large quantities o
sea water [2]. The third step is to convert the CO_2 and
fresh water to syngas (carbon monoxide (CO) and
hydrogen gas (H_2)) [3}. The fourth and final step is to
convert the syngas via the Fisher-Tropsch synthesis to
petrol (gasoline) and diesel fuel [4}.These CO_2
removal/desalination/syngas production, and synthetic
petrol production plants would best be located around
sea coasts. The Synthetic petrol and diesel from air and

water would make the process very economical.

References

[1] Klaus S. Lackner (Columbia University), Patrick Grimes (Grimes Associates), and Hans-J. Ziock (Los Alamos National Laboratory), "Capturing Carbon Dioxide From Air," available at http://www.inl.gov/featurestories/docs/parametric_study of_large-scale_production_of_syngas_via_high_ temperature_co-electrolysis.pdf

[2] General Electric's Desalination Process, available at

http://www.gewater.com/what_we_do/water_ scarcity/desalination.jsp

[3] J. E. O'Brien, M. G. McKellar, C. M. Stoots, J. S. Herring, and G. L. Hawkes (Idaho National Laboratory), "Parametric Study of Large-Scale Production of Syngas via High Temperature Co-Electrolysis," available at http://www.inl.gov/featurestories/docs/parametric_ study_of_large scale_production_of_syngas_via_high _temperature_ co-electrolysis.pdf

[4] Tyler Hamilton, "Natural Gas to Gasoline: A firm claims to have a cheaper way to harness natural gas," Technology Review, August 15, 2008, available at http://www.technologyreview.com/energy/21261/?a=f

Chapter 9
Concluding Letter-to-the-Editor

The Saint Louis Post-Dispatch has run disparaging pieces about biofuels, especially ethanol. Much opinion about biofuels is based on incorrect information. Now, with global warming on the rise, comes new information about biofuels that makes them much more attractive.

Greenhouse gas emissions may be drastically reduced through biofuel production. "Challenges to Scaling up Biofuels Infrastructure," an article by Tom L. Richard in Science in August, said:

"The next few decades will require massive growth of the bioenergy industry to address societal demands to reduce net carbon emissions. This is particularly true for liquid transportation fuels, where other renewable alternatives to biofuels appear decades away, especially for truck, marine, and aviation fuels. But even for electricity and power, the growth potential for other renewables and nuclear power appears limited by high cost, technology barriers, and/or resource constraints."

The magazine said that with estimated bioenergy potential ranging from slightly less than 10 percent to more than 60 percent of "world primary energy, biomass seems poised to provide a major alternative to fossil fuels."

It also said , "As a point of reference for considering future biomass infrastructure needs, the International Energy Agency estimates" a 50 percent reduction in greenhouse gas emissions by 2050 will require bioenergy production to quadruple.

Clearly, we need to move forward on biofuels.

Gregory Gebhart • Webster Groves

Chapter 10
Shale Oil and Natural Gas

The American oil and natural gas industry is trumpeting the advantages to shale oil and natural gas. Unfortunately both are incredibly harmful to the environment. To produce Shale Oil, billions of gallons of fresh water need to be boiled with ground up shale oil rock. This irreversible pollutes billions of gallons of fresh water when the U.S. is running short of water in the western half of the U.S. And natural gas is no better. Fracking the underground shale deposits with high pressure toxic chemicals is polluting billions of gallons of underground freshwater aquifers in the eastern half of the U.S. Biofuels are the future. Oil and natural gas are the past.

Bibliography

Agricultural Statistics 2004, U.S. Department of Agriculture, Tables 1-4 (Wheat: Area, yield and production by State, 2001-2003), 1-7 (Wheat: Supply and disappearance, United States, 1994-2003), 1-37 (Corn: Area, yield and production, by State, 2001-2003), 1-38 (Wheat: Supply and disappearance, United States, 1994-2003), 1-39 (Corn: Utilization for silage, by State, 2001-2003), 1-40 (Corn for grain: Marketing year average price and value, by State, crop of 2001, 2002, and 2003), 1-41 (Corn: Area, yield, and production in specified countries, 2000/2001-2002/2003), 1-42 (Corn: International trade, 2001-2003), 1-43 (Corn: Support operations, United States, 1994-2003), 1-44 (Corn: United States Exports, specified by country of destination, 2000/2001-2002/2003), 1-45 (Oats: Area, yield, production and value, United States, 1994-2003), 1-46 (Oats: Stocks on and off farms, United States, 1994-2003), Table 1-47 (Oats: Supply and disappearance, United States, 1994-2003), 1-49 (Oats: Support Operations, United States, 1994-2003), 1-49 (Oats: Area, yield, and production, by State, 2001-2003), 1-50 (Oats: Marketing year average price and value, by State, crop of 2001, 2002, and 2003), 1-51 (Oats: Area, yield, and production, in specified countries, 2000/20001-2002/2003), 1-52 (Barley: Area, yield, production and value, United States, 1994-2003), 1-53 (Barley: Stocks on and off farms, United States, 1994-2003), Table 1-54 (Barley: Supply and disappearance, United States, 1994-2003), 1-55 (Barley: Area, yield, and production, by State, 2001-2003), 1-56 (Barley: Marketing year average price and value, by State, crop of 2001, 2002, and 2003), 1-57 (Barley: Area, yield, and production, in specified countries, 2000/20001-2002/2003),

1-58 (Grains and grain products: Total per capita civilian consumption as food, United States, 1993-2003), 1-59 (Barley: Support Operations, United States, 1994-2003), 9-1 (Economic Trends: Data relating to agriculture, United States, 1993-2003), 9-2 (Farms: Number, land in farms, and average size of farms, United States, 1994-2003), 9-3 (Farms: Percent of farms, land in farms, and average size, by economic sales class, United States, 2002-2003), 9-4 (Number of Farms: Economic sales class, by region and United States, 2001-2003), 9-5 (Land in farms: Economic sales class, by region and United States, 2001-2003), 9-6 (Land in Farms: Classification by tenure of operator, United States, 1910-2002), 9-7 (Farms: Classification by tenure of operator, United States, 1915-2002), 9-8 (Farms: Classification by Tenants and Part Owners, United States, 1900-1997), 9-9 (Farms: Number and land in farms, by State, 2001, 2002, 2003), 9-10 (Land: Utilization by States, 1997), 9-11(Land in farms: irrigated land, by State, 1959-1997), 9-12 (Farm real estate: Value of farmland and buildings, by State, 1999-2003), 9-13 (Land utilization, United States, selected years, 1940-1997(, 9-14 (Farm real estate: Average value per acre of land and buildings, by State, Mar 1, 1970 and Jan 1, 1999-2003), 9-15 (Land values, cropland and pasture: By State, 2002-2003), 9-16)Cash rents, cropland and pasture: By State, 2002-2003)

Bartoszek-Loza, Rosemary, "Novel Catalysts and Processes for the Photochemical Decarboxylation of Alpha-Hydroxy-Carboxylic Acids," U.S. Patent # 4,515,667, issued May 7, 1985 (Assignee: The Standard Oil Company [Cleveland, Ohio])

Marco a. van den Berg, Patricia de Jong-Gubbels, Christine J. Kortland, Johannes P. van Dijken, Jack T. Pronk, and H. Yde Steensma (Institute of Molecular Plant Sciences, Lieden

University, Lieden, The Netherlands), "The Two Acetyl-coenzyme A Synthetases of Saccharomyces cerevisiae Differ with Respect to Kinetic Properties and Transcriptional Regulation," Journal of Biological Chemistry (The American Society for Biochemistry and Molecular biology), Volume 271, Number 46, Issue of November 15, 1996, pages 28953-28959

G.E. Bullock, Ethanol from Sugarcane, Sugar Research Institute for Queensland Department of State Development within the over-arching program of projects titled "Sugar Industry Renewable energy", December 2002

DeWitte, Justin E., Prevost, John M., Li, Zhong, Krausman, Howard W. (Assignee: DaimlerChrysler Corporation), "Engine Operations in an unknown ethanol blend," U.S. Patent # 6,851,416, issued February 8, 2005

Christina Galitsky, Ernst Worrell, and Michael Ruth, "Energy Efficiency Improvement and Cost Saving Opportunities for the Corn Wet Milling Industry", "An ENERGY STAR Guide for Energy and Plant Managers" (Environmental Energy Technologies Division, Ernest Orlando Lawrence Berkley National Laboratory Report sponsored by the U.S. Environmental Protection Agency) July 2003

Grove, Leslie H. "Process for the Production of Organic Fuel", U.S. Patent # 4,326,032, issued April 20, 1982

<http://www.bbiethanol.com/grain/>
<http://www.bbiethanol.com/biomass>
http://www.bbiethanol.com/plant_production/usnc.html
<http://bioenergy.ornl.gov/papers/misc/energy_conv.htnl>
<http://www.consumerenergycenter.com/transportation/afv/e

thanol.html>
<http://www.ford.com/en/vehicles/SpecialtyVehicles/environ
mental/ethanol.html>
<http://www.fleet.ford.com/showroom/2005fleetshowroom/2
005-taurusFFV>
<http://www.fleet.ford.com/showroom/2005fleetshowroom/2
005-explorerFFV>
<http://www.fleet.ford.com/showroom/2005fleetshowroom/2
005-mountainerFFV>

Min Liu, Nanqi Ren, Jie Ding, Peng Li(School of Municipal
& Environmental Engineering, Harbin Institute of
Technology, Harbin, People's Republic of China), "The
Effect of Organic Loading Rate and pH on Bio-Hydrogen
Production from Starch Wastewater in CSTR (Continuous
Stirring and Temperature Reactor),
<http://waterstof.org/20030805EHECO-57.pdf>

Margarida Moreira dos Santos, Andreas Karoly Gombert,
Bjarke Christensen, Lisbeth Olsson and Jens Nielsen (Center
for Process Biotechnology, BioCentrum-DDU, Technical
University of Denmark, Lyngby, Denmark), "Identification
of In Vivo Enzyme Activities in the Cometabolism of
Glucose and Acetate by Saccharomyces cerevisiae by Using
13C-Labeled substrates," Eukaryotic Cell, June 2003,
Volume 2, number 3, Pages 599-608

Jonathan Scurlock (Oak Ridge National Laboratory,
Bioenergy Feedstock Development Programs (Oak Ridge,
TN) managed by University of Tennessee-Battelle, LLC for
the U.S. Department of Energy), "Bioenergy Feedstock
Characteristics",
<http://www.ott.doe.gov/biofuels/properties_database.html>

Hosein Shapouri (U.S. Department of Agriculture (USDA),

Office of the Chief Economist) an Andrew McAloon (USDA Agricultural Research Service, Eastern Regional Research Center)"The 2001 Net Energy Balance of Corn-Ethanol," updated March 22, 2004

Hosein Shapouri, James A. Duffield, and Michael S. Graboski (U.S. Department of Agriculture, Economic Research Services, Office of Energy), "Estimating the Net Energy Balance of Corn Ethanol, Agricultural Economic Report number 721, July 1995

Takahiro Suzaki, Takeshi Matsuo, Kazuhisa Ohtaguchi, and Kozo Koide (Department of Chemical Engineering, Tokyo Institute of Technology, Tokyo, Japan), "Kinetics of Production of Acetic Acid from Carbon Dioxide by Acetobacterium woodii in Bubble-Column Bioreactor," in Repeated-Batch Cultures using Flocculated Cells of Acetobacterium woodii," Journal of Chemical Engineering of Japan, Volume 25, Number 1, Pages 106-108 (1992)

Takahiro Suzaki, Takeshi Matsuo, Kazuhisa Ohtaguchi, and Kozo Koide (Department of Chemical Engineering, Tokyo Institute of Technology, Tokyo, Japan), "Continuous Production of Acetic Acid from CO2 in Repeated-Batch Cultures using Flocculated Cells of Acetobacterium woodii," Journal of Chemical Engineering of Japan, Volume 26, Number 5, Pages 459-462 (1993)

Marcia Sadae Tano and Joao Batista Buzato (Departmento de Bioquimica, Universidad Estadual de Londrina, Londrina, Brasil), "Effect of the Presence of Initial Ethanol on Ethanol Production in Sugar Cane Juice by Zymomonas mobilis," Brazilian Journal of Microbiology, Volume 34, Number 3, Jul/September 2003

R.F. Tester, J. Karkalas, and X Qi, (Department of Biological Sciences, Glascow Caledonian University, Glasgow, UK), "Starch Structure and Digestibility Enzyme-Substrate Relationship," World's Poultry Science Journal, Vol 60, June 2004, Pages 186-195

Kaimal, Thengumpillil; Narayana Balagopala; Vijayalakshmi, Penumarthy; Ramalinga, Bandi, Laxmi; Ayyagari Ananta (all of Pradesh, India), "Process for the Preparation of Alkyl Esters from Commercial Lactic Acid", U.S. Patent # 6,342,626, issued January 29, 2002 (Assignee: Council of Scientific & Industrial Research, New Delhi, India)

U.S. Public Interest Research Group, "Global Warming," <http://uspirg.org/uspirg.asp?id2=5235&id3=USPIRG>

Verser, Dan and Eggeman, Tim (Assignee: ZeaChem, Inc.), "Process for producing ethanol," U.S. Patent # 6,509,180, issued January 21, 2003

Robert Wallace, Kelly Ibsen (National Renewable Energy Laboratory, National Bioenergy Center), Andrew McAloon, Winnie Yee (U.S. Department of Agriculture, Eastern Regional Research Center, Agriculture Research Service), "Feasibility Study for Co-Locating and Integrating Ethanol Plants from Corn starch and Lignocellulosic Feedstocks," A joint Study Sponsored by U.S. Department of Agriculture and U.S. Department of Energy, <http://www.osti.gov/bridge>, January 2005

www.ingramcontent.com/pod-product-compliance
Lightning Source LLC
Chambersburg PA
CBHW031236280526
45784CB00004B/1599